Paul Mobbs

Energy Beyond Oil

This book has been produced to document two years of work as part of the "Energy Beyond Oil" project. For further information about the project, free information, and material produced to accompany this book, go to the Energy Beyond Oil web site: http://www.fraw.org.uk/ebo/

Matador Publishing
9 De Montfort Mews
Leicester LE1 7FW, UK
Email: books@troubador.co.uk
Web: www.troubador.co.uk/matador

ISBN 1 905237 00 6

Printed on 100% post-consumer recycled paper

Typeset in 11pt Gill Sans by
Troubador Publishing Ltd, Leicester, UK

Printed by The Cromwell Press Ltd, Trowbridge, Wilts, UK

Matador is an imprint of Troubador Publishing Ltd

Contents

List of Information Boxes, Figures and Tables

Introduction

Since 2002 I've been holding a discussion...

I've had this discussion in the street, on trains, on the phone, at parties, and in many, many meetings. It involves a fairly simple argument that relates to the fields of chemistry, physics, economics and the environment. The discussion is a mechanistic sequence of arguments and interpretations of data produced by governments and international agencies. It often results in the participants in the discussion either stuck for words, dismissive of the data, or they describe the interpretation as environmental extremism. I've written this book in order to share this discussion more widely.

The discussion usually runs along the following lines (the participant's general responses are in italics):

There is a fundamental physical law of the universe – The First Law of Thermodynamics. It states:

"in any closed system, the total amount of energy of all kinds is a constant"

The Earth is a closed system. It receives energy from the Sun, and the Moon pulls the oceans across the globe to make the tides. Apart from these two, almost constant, energy inputs from the outside the Earth has no other external energy sources. Any other energy source we use has to come from the Earth itself... and these sources are running out.

Oil is almost at peak production – after that, it's on its way out. Within ten to twelve years, when the pumps can suck no greater volumes of oil from the ground, production will go into decline and it will never rise again. Oil deposits will be exhausted, in terms of oil being a bulk energy resource, around 2050.

So what's the problem, it going to last another fifty years?

Market economics... When oil production reaches its peak, from that day on every state across the globe will be competing on price for the ever-dwindling level of production that remains. Result: between 2010 and 2015 the price of oil will start climbing higher and higher, doubling or tripling in price in just the first ten to fifteen years after the peak, and, from that point on, it will never fall in price again. We could shift to gas, but for many of the uses of oil – such as a fuel in cars or as the feed material for the production of plastics – natural gas just can't provide as efficient a replacement.

So what?... all those renewable technologies will get comparatively cheaper?

Energy density... Oil and gas are actually very dense sources of energy –

they contain a large amount of energy per unit of mass compared to other sources. Let's compare motor fuel with its nearest renewable equivalent, biodiesel. It takes two hundred and fifty gallons of diesel per year to keep the average diesel car going the average distance travelled every year. Replace that with biodiesel, and you have to find a hectare ($2^1/_2$ acres) of land to produce the three tonnes of oilseeds that biodiesel production requires. Want to make the twenty-odd million the cars in the UK run on biodiesel? – sorry, but that would take five times more land than all the farm land currently in cultivation in the UK. Even if we turned over half the cultivated land in the UK to produce biofuels it would only keep just over two million cars (less than 10% of the current car fleet) on the road – and of course this figure doesn't include lorries, trains, tractors, etc., which would also demand biodiesel to keep running.

But what about hydrogen, that wonder-fuel of the future?

You need energy to make it... The Earth is a closed system and we don't have natural hydrogen reserves. We can't magic it from the air, water, or any other source. Hydrogen itself is not a fuel. Like electricity it is a carrier of energy. We have to put the energy into processes that makes hydrogen in order to produce a fuel that gives us back only 40% to 80% of the energy we invested in making it. Currently hydrogen production relies on the use of hydrocarbons like oil and gas. Of course, there is also the slight problem that large-scale hydrogen use could make a huge hole in the ozone layer every bit as bad as all those chlorofluorocarbons that we banned fifteen years ago.

OK, if we're desperate we can go nuclear can't we?

There's not enough uranium... Even if we ignored all the problems with nuclear power [for example, no one has yet found a fully reliable method for storing the highly radioactive wastes produced for a hundred thousand years] if the world switched most of its electrical requirements to nuclear, there's only enough uranium to keep fission reactors burning for fifteen to twenty years. We could try and get the more risky fast breeder reactors working reliably (not an easy job) to make the uranium go further... in which case we might get fifty or sixty years of energy. The only longer-term nuclear option is fusion. But that's perhaps a century away, and, even then, it's not clear that it could provide for the current energy consumption of the globe for generations to come.

But I thought we could have more oil from Iraq now?

Iraq's oil represents just 4 years global use... Iraq's proven oil reserves are 112.5 billion barrels – global consumption in 2004 was 29 billion barrels, and that's rising two to four percent per year driven by the industrialisation of India and China. New oil and gas finds are getting smaller, and are not replacing current consumption. So the total level of oil and gas resources is falling. Also, as existing oil fields reach about half of their viable production, the level of output begins to fall off because it's harder to suck the oil out. Put this together, and Iraq really doesn't make a lot of difference given the scale of global consumption.

Well, we've still got two hundred years worth of coal!

Climate change makes its use impossible... We might have had two hundred years worth in 1950, but at today's level of electricity consumption the UK's 1.5 billion tonnes of coal reserves will only last us nine years. Coal can be turned into oil and gas and all sorts of materials. But coal isn't as efficient to use to produce energy as oil and gas because it mainly consists of carbon, not hydrocarbons. Therefore you have to burn more of it to get the same energy output, and you'd still have problems making other products like fertilisers or plastics. Most significantly, if the world used coal to replace oil and gas the increase in carbon emissions would mean that climate change might wipe us out.

Expensive oil, renewables are not enough, no coal, no nukes... we're in trouble!

Yes, and that starts in the next five to ten years when energy prices rocket – not in fifty years when oil is scarce.

It's at this point that many participants begin to experience what could be described as metanoic shock. Old certainties don't seem so certain any more. And the comfort of our Western lifestyle suddenly seems to represent a thin veneer over the twilight zone where cheap holidays and off-road vehicles are as fond a memory as steam trains.

But this is what we face. It's the laws of physics. You can't create something out of nothing. More interestingly, why are politicians not discussing this, given we're within about two average political terms of when these problems start? That's not clear. It might be that the economists' faith in the market, and that the market will always provide a solution, might be blinkering the view politicians take on the larger issue of energy supply. Perhaps they believe that there will always be alternatives, such as nuclear or renewable sources, that could substitute for our diminishing sources of energy (however, anyone who believes this clearly hasn't done the maths). But there is one other possible answer – since when do politicians promise you less?

What's heartening is that some people see this impending realignment of global energy quotas as a great opportunity. A chance to redefine the balance between humans and the planet in order to create a sustainable future. Of course, many of these people, like myself, were brought up to cook their own food from raw ingredients, to save and mend, and to generally shun the consumer lifestyle. I find that those who are extremely hostile to the arguments over future energy shortages usually represent the antithesis of these traits.

And herein lies the problem. There are many other people talking about energy shortages and *Peak Oil*, mainly in the USA, but usually their discussion is based on how to keep their living standards at their current level. This, of course, is an impossibility. It's that horrid *First Law of Thermodynamics* – if we reduce the level of energy within the system the level of activity inside the system must also contract.

So, here we are. Whilst you read this sentence the world, on average, has just burnt another seven to eight thousand barrels of oil. In fact, it gets through around eighty-two million barrels per day. In order that you can share the argument let's explore the issues in more detail. The data. The trends. The projections. The possible outcomes. Hopefully at the end of this process you will be able to understand what it is we are facing, and perhaps find your own resolution to the potential difficulties we will all face over the next ten to twenty years.

I hope that the message you take from this book is a positive one. That Western society is about to undergo a massive, collective shock, but, by applying basic principles of sustainable development we can live through this period... albeit without the ready-meals, cheap flights to Spain, 4x4's, Britney Spears videos, Formula One racing, plastic umbrellas...

Paul Mobbs
February 2005

1. What Is It About Oil?

Energy. Seems so simple... *it's just, there.*

Perhaps this is the problem. It's difficult for people to imagine the scale of the problems facing industrial society due to the imminent decline of fossil fuels because energy has always, in the industrialised countries at least, seemed so plentiful. For many, it is also difficult to appreciate just how much energy the world uses.

There is another way of interpreting our scant concern for energy resources. Energy seems so simple because the public has never been encouraged to think about it, but merely to consume. And in fact, if you do try and look at the bigger picture, things will be made very difficult for you because of the way governments and energy experts present information about our use of energy. For example, a large part of the public discussion of energy relates to domestic energy consumption, when in reality domestic uses only make up 30% of the UK's total energy consumption.

This book is in part about oil, and the way oil is at the heart of energy use in developed world states like the UK. However, to assess the value of the energy oil gives to society, and what happens when it goes into decline, we have to look at the comparative value of other energy sources too. Common sense dictates that if we're talking about the absence of oil we have to understand the relationship between how oil is used and how other energy sources might replace that energy. More importantly, as the energy contained in oil isn't directly comparable to the energy produced by gas, or nuclear power, or renewable energy, we need to understand how we might efficiently convert the value of one energy source into another.

Other problems arise when we look at energy consumption overall. We have to be able to convert all the different sources of energy into a common unit of measurement. If we factor in other details, such as the efficiency of how we use energy when we switch fuel sources, this complicates things further. How we examine and measure our use of energy will be dealt with throughout the book. But to begin, just what is it about oil that makes it so important?

Oil, gas and coal are mineral resources, like many of the other mineral resources that we mine or process every day to produce everything from metals to cat litter. What distinguishes oil, gas and coal is that they contain a lot of energy, and for that reason they have become far more important to society than many of the other commodities produced from minerals. But it is *how* we use energy that is of greater relevance than the fact that oil and gas give us a plentiful energy supply. This is especially true given that the majority of our current energy sources will not be available to use in the foreseeable future. Either because (like oil) they will run out, or become very expensive, or because (like coal) their use creates other unwelcome consequences such as climate change. By understanding the physical processes of how energy can be sourced, stored and moved, we can assess what alternative sources would be credible replacements when fossil fuels can no longer deliver the energy we need.

Although coal and gas supply a lot of our energy today, more than anything it is oil that makes the world go around. It doesn't just provide a lot of the energy for our society. It also oils the bearings, holds the roads together, and is prime ingredient in many of the materials that make up the world around us. It's just so... *useful.* So the reduction in oil supply doesn't just create a shortage of petrol or heating oil. It has knock-on effects throughout the entire commodity supply chain of the developed world.

It is a fact that the oil is running out. It's fairly simple mathematics. Each year a few small oil fields are discovered adding to the global reserves. In that same year a larger volume is pumped from the ground and used by industrial society. Consequence: the global reserves of oil are falling year on year. Most people in the Western world are tacitly aware of this. They are also aware that it may be some decades before the last oil well runs dry, and so it's not an immediate problem. This ignores an important factor that most people within the energy industries, and within the economics departments of global institutions and national governments, are far more concerned about. Long before the date when the oil runs out, the price of oil and gas is going to rise, and after this point in time the price will never come down again.

Peak Oil is the name given to the date when the world simply can't pump any more oil. The difficulty we have with the timing of Peak Oil is that the nebulous predictions of geophysical science give a range of possible dates for when it will occur. Most are in the range of 2005 to 2015, although some of the more optimistic studies place it between 2020 and 2035. Even so, international think tanks are writing papers on it [Mobbs, 2004a], and some large corporations are worried about it. It just seems that the message hasn't percolated down to the public so that it can form a topic of popular discussion. There is also a clear silence within the political arena, but this might be due to the fact that the results of Peak Oil could be rather unsettling to the traditional political message that *things can only get better* [Guardian, 2004a].

Oil is one of the most important trading commodities. The fact that people demand so much of it means that in times of war or political uncertainty the price of oil will rise across the world as corporations and governments buy in more stocks to guard against future uncertainties over price or the level of supply. Whilst short-term political factors may affect the oil price over a few weeks, it is the availability of oil through the purchasing of supply contracts that determines the longer-term trend in oil prices.

Due to the trends in oil consumption, especially the large demand from China and India, during 2003 oil prices rose because higher demand was applying pressure for oil supply to be increased. Consequently, during the Spring and Autumn of 2004, prices rose to the highest dollar value in history (although the actual real terms value, taking into account inflation, was only half that of the record prices of 1980). In the latter half of 2004 oil prices began to fall as the Organisation of Petroleum Exporting Countries (OPEC) signalled an increase in oil supply. However, what if OPEC could not promise such an increase? This is the simplest meaning of Peak Oil – the world's oil producers will not be able to sustain the historical increases in world oil production.

Within the next five to ten years we may reach and pass the date of Peak Oil. In the few years following the date of Peak Oil the global supply of oil must fall, but it is certain that the demand for oil will remain strong. So as night follows day, it will mean that global commodities brokers will be in a bidding war to secure the rights to the remaining sources of oil. The price will go up, and up, and how far it rises will be related to how the world's political leaders react to the growing energy crisis. This is another significant issue related to Peak Oil. It doesn't matter how good the technological systems are to replace oil, if the world's political leaders can't deal with these changes amicably then we're in for a really bad time.

It's hard to overestimate the importance of oil to industrialised society. It isn't just an energy source. It's much more than that. The products of oil and gas, such as plastics and agricultural fertilisers, and a large part of the world's chemical industries and manufacturing processes, are directly dependent upon the refining of crude oil and natural gas. As an energy source oil is also unique. It's not just that oil is a very dense source of energy – it packs a lot of energy into a small space. It also just happens to be a liquid at the Earth's ambient temperature. That makes it perfect for use in mobile sources. You don't have to use special pressurised storage systems like you do for gas. You don't have to have complex handling systems like you have to have for solid or particulate materials like coal. It's a liquid – you can pour it, pump it, and you can can utilise it without putting a lot of energy into the process that produces the required energy output.

The suitability of oil for use in mobile sources is key to understanding the significance of oil to the economy. The depletion of oil resources isn't just about replacing energy. It's the physical nature of oil that's as important as the energy it contains. You can have battery powered vehicles but carrying and charging batteries uses a lot of energy in itself, adding to the burden of energy demand. You can have hydrogen or gas powered vehicles but they still need energy or fuel to operate and move around. So oil isn't just a source of energy. It's a fuel that is important because of its physical form. What's also important is that when the refining process turns crude oil into a liquid fuel it produces many 'waste' products which are the primary feedstock of a large part of the globe's industrial capacity – from tarmac, to plastics and paints.

All life on Earth sources its energy from the Sun. The Sun shines at an almost constant rate, and so each year the plants and animals on the Earth can harvest a constant amount of energy from the environment around them. Oil contains a fraction of the energy that ancient organisms absorbed from the Sun millions of years ago. Each oil field represents, in a concentrated form, the energy that was absorbed by plants and animals over tens or hundreds of thousands of years. By extracting the oil and burning it we're cheating the solar input to the Earth. We're getting more out than the Sun is putting in at the present time. Consequently not only are we emptying the Earth's piggy bank of energy, but when it's gone society must undergo an energy shock as the lack of energy forces a contraction of economic activity. Ultimately, when the oil, gas or even coal runs out, we will have to return to using solar energy again – either directly, or as plant matter (usually called biomass).

The organic matter that produced the oil is high in the chemicals that make up life, most significantly carbon. This organic matter has been turned, by vast geological processes, into very useful chemical compounds that are the basis of a large part of our manufacturing industries. We can reproduce those compounds by growing more plants today but we are limited by the area of land that is required to produce a certain level of materials output. In fact, there just isn't enough land to grow these crops intensively to replicate the levels of energy and chemical inputs oil represents to society. Not just because of the land take to produce the same volume of output, but because at the same time the effects of climate change will diminish the areas of viable agricultural land across the globe. So by losing oil we don't just lose energy – we lose a large proportion of those other products that we make from oil.

Where does oil come from? Yes, *the ground*, but it's important to consider what it's made from, and why it is so important to modern society.

Oil is the product of *fossil biomass*. It's the sticky ooze that remains from ancient plants and animals after they have been compressed and heated by geological processes for millions of years. Science is still not entirely certain how oil is produced. We know the general conditions for its formation, but not the precise sequence of chemical processes that turned plants into hydrocarbons.

Recently some people have even speculated that oil doesn't come from buried organisms, but instead it wells up from the Earth's upper mantle where an infinite oil field exists that was created when our planet was made. This so called *abiotic theory of oil formation* has next to no evidence to support it, and runs counter to the geophysical principles that have underpinned the development of modern Earth science. However it is being used by those who oppose the concept of Peak Oil, or who oppose the development of renewable energy systems, to argue that the oil isn't going to run out because if we wait long enough the oil wells will refill themselves. This distortion of science by those who oppose the radical changes required by Peak Oil, or climate change, is discussed later in the book.

The scientific consensus across the globe is that oil originates from aquatic plants and animals. These die and fall to the bottom of the sea. What's important to this process is that the sea bed contains very little free oxygen. This prevents sea-bed scavengers from eating the organic matter and breaking it down too soon. As the organic-rich sediments are buried deeper and deeper, to a depth of thousands of metres, the increasing temperature and pressure slowly turn the organic matter into oil and gas. Under similar conditions, where the organic material is mainly composed of woody plants, you get coal.

Most of the world's oil comes from *primary deposits*. This consists of free oil held in underground reservoirs created by geological processes (see Box 1). Other geological conditions will also lead to the trapping of oil, but these *secondary deposits* tend to be of lower quality than the conventional oil deposits trapped in reservoir-like geological formations. For example, if the formation of oil takes place within bands of impermeable silts and muds then the oil is not able move. It stays where it was created. This produces what's called *oil shale*. The oil can be extracted from the shale, but it leaves large quantities of rock that has to be disposed of.

The other main sources are *heavy oil deposits* and *tar sands*. These are deposits that are high in tar, and so the deposit itself prevents the movement of the lighter fraction of the oil. Heavy oil can be extracted by piping high temperature steam into the deposit to boil off the lighter fraction, but the extraction wells produce a lower quality product. For both oil shales and heavy oil, the recovery process itself requires energy and so the net energy output (output minus the energy input to the process) is lower than for primary oil sources.

From 1810 to 1820, many European Countries started to produce oil and gas from coal by *carbonisation*. Coal was heated in a sealed vessel until it broke down producing town gas, gas oil, and coke. Around the same time as the carbonisation of coal began small volumes of crude oil had been discovered and

5

Box 1. Oil Bearing Structures

The creation of oil (and gas) reservoirs is entirely dependent upon the structure of the rocks around the area where the oil was formed. Oil is less dense than water – as you might have noticed it floats on the surface of water because it is lighter. Underground rocks, particularly those formed from sediments, are saturated with water. Depending on how much sand or limestone is in the rock, the rock will be permeable, and both water and oil will flow though it. So as the oil is formed in the water saturated rock it becomes buoyant and begins to float upward.

Layers of rock can be folded and fractured by geological processes. Where a permeable rock contains material that forms oil and gas these geological structures can trap oil and gas to create a reservoir. Figure 1 shows an *anticline reservoir* – a fold in the rock strata. This type of structure holds 80% of the world's oil reserves. It also shows a *faulted reservoir* where the displacement of the rock on one side of the fault means that the oil bearing rock abuts a layer of impermeable rock. This forms a trap, allowing oil and gas reservoirs to form.

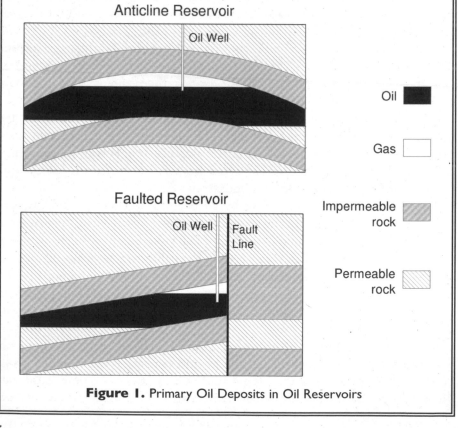

Figure 1. Primary Oil Deposits in Oil Reservoirs

collected from coal mines. Some experiments were carried out to refine this oil, but it could not compete on volume with the production of gas oil from town gas works. In Scotland, from the 1860s, oil shales were quarried to produce petroleum. This process operated until the early 1970s when it was made redundant by the use of natural gas from the North Sea. It was the use of compounds created from the products of coal carbonisation, and early oil refining, that laid the foundation of the modern petrochemical industry.

It was the improvements in geological science that led to the discovery of the world's major oil fields. In some areas of the world oil reservoirs extended to the surface, forming pools of oil or tar pits. By identifying the geological structures that held crude oil geologists were able to explain how oil reservoirs were formed, and were able to speculate where oil might collect. In this way prospectors were able to search for oil by drilling into the types of geological formation that were most likely to contain oil-bearing rock strata. This informed approach yielded the large finds in North America and the Middle East at the end of the Nineteenth Century.

Following World War Two, new technology changed the approach of prospectors. *Geophysics*, the ability to undertake experiments at ground level to determine the structure of the rocks beneath, changed the way that the industry looked for oil. Now it was possible to survey an area first, to identify structures deep below the ground most likely to hold oil, before sinking wells precisely into the best rock formations. Over the last forty years, using geophysics, the oil companies have explored just about the entire globe looking for oil – either on land, or under the sea on the continental shelves (oil does not form in the deep oceans). As exploration was carried out there were major finds, such as the North Sea, and more recently in Latin America, West Africa and South East Asia. However none of these, or more recent, finds, are anything like the scale of those found in North America, the Middle East, Russia and some Central Asian states. The fact that finds are getting smaller also means that oil production companies cherry pick the best production sites. The effect of this is to exhaust the best reserves first, making it progressively harder to produce oil from the remaining reserves, and in the longer-term this will raise the costs and reduce the level of oil production.

Both oil and gas are *hydrocarbons*. The structure of a number of hydrocarbons are shown in figure 2. Hydrocarbons are made from molecules containing carbon atoms (abbreviated C) with hydrogen atoms (abbreviated H) bonded to them. When oil or gas are burnt – that is, reacted with oxygen (abbreviated O) – the chemical bonds of the hydrocarbon chains are broken, producing energy.

In an ideal world, for example when burning methane, this produces carbon dioxide (CO_2, one carbon atom bonded to two oxygen atoms) and water

(H₂O, two hydrogen atoms bonded to one oxygen atom). In the real world combustion isn't so efficient so you also get carbon monoxide (CO) and other compounds based on nitrogen as the nitrogen in the air reacts at high temperatures.

   Producing oil for burning in engines is not the only purpose of mining crude oil. Today an equally significant use is the production of chemical compounds and hydrocarbon products. This requires that the precise constituents of the crude oil are matched to its final use.

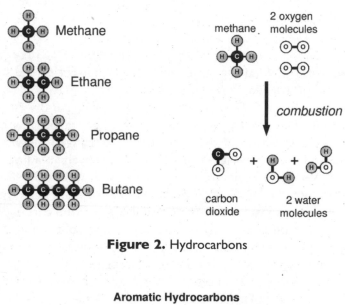

Figure 2. Hydrocarbons

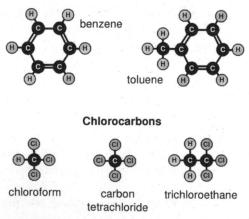

Figure 3. Complex Hydrocarbons

Not all crude oil is the same. There are different types of oil, and different oil reserves produce different blends of oil. *Sweet* or *light crude* is found in the North Sea and North Africa, and has higher concentrations of short-chain hydrocarbons that are easier to refine into petrol. *Heavy crude* is found in the Middle East and other locations around the world such as Venezuela, and has higher concentrations of long-chain hydrocarbons. These require more refining to produce petroleum products, but can produce other hydrocarbon products such as plastics or road tar.

In addition to hydrocarbon chains, the other significant constituent of crude oil are *aromatic hydrocarbons*. These are complex structures created with rings of six carbon atoms, hydrogen and other elements such as oxygen and nitrogen (see figure 3). Aromatic hydrocarbons are very important as a feedstock for the chemical industry.

Different types of crude oil will have varying quantities of aromatic hydrocarbons present. So, depending upon whether the oil is destined for refining into fuel, or for use in the chemical industry, oil from different regions of the world will be blended to produce different classes of crude. Finding the right types of oil for blending will be one of the early problems to affect the petrochemical industry after the peak in oil production. Although these different carbon structures can be made in refineries, this will use up more energy – more oil – compared to harvesting these compounds directly from crude oil.

When crude oil is refined the different weights of carbon compound, determined by the length of the carbon chain, are separated out. Different hydrocarbons condense from a gas to a liquid at a different temperatures. By heating the crude oil to between 400°C and 600°C, and letting the gases cool in a tower-like structure called a *condenser*, it is possible to extract a relatively pure stream of hydrocarbons of a certain weight by trapping the condensate at different temperature zones. This process is called *fractional distillation*.

Some of the compounds liberated when the oil is heated will not condense at ambient temperatures – they are gases such as methane, ethane, butane and propane. These can be used as a fuel directly (called *liquid petroleum gas*, or *LPG*), or they can be purified and bottled (for example, butane and propane), or used as the feedstock for chemical processes (for example, methane and ethane). The *light distillate*, that condenses at 70°C to 200°C, produces fuel oils such as petrol and kerosene, as well as light hydrocarbons for use in the chemical industry. The *middle distillate*, that condenses at 200°C to 350°C, produces heavier fuel oils like diesel and gas oil (used for fuelling industrial boilers). The *heavy distillate*, that condenses above 350°C, produces bunker oil for marine engines and oil-fired power stations, as well as heavy tars for the production of chemicals and for use in products such as roofing bitumen. What's left – that is, what didn't boil and produce gas – is made from compounds containing very dense forms of carbon. This too can be used as the feedstock for chemical processes, or as the basis for producing the road tar used to make tarmac.

Of course, it would be rather a waste to reject the heavier fractions produced by distillation given a large part of the demand for hydrocarbons is as light fuel oil. So processes have been developed to take the heavier fraction of distillate and process it to produce lighter ones, often for use by the chemical industry. This uses a process called *cracking*. Using heat, pressure, and catalysts that aid chemical reactions, the long hydrocarbon chains are broken into smaller sections. The shorter hydrocarbons produced are then refined again in order to separate the different fractions.

As well as breaking hydrocarbons apart it's also possible to put them together. This is how plastics, or *polymers*, are produced – illustrated in figure 4. Both distillation and cracking produce very short chain hydrocarbons. One of these, ethane, can be processed to produce ethylene. Ethylene is known as a *monomer*, because it is used as the building-block for producing larger molecules by joining lots of monomer molecules together in a process called *addition polymerisation*. When using ethylene, polymerisation produces a huge macromolecule, *polyethylene*, commonly called polythene. Other more complex reactions, called *condensation polymerisation*, use a variety of hydrocarbon compounds to produce other plastics such as *polyamide*, commonly known as nylon, and *polyester*. By removing the hydrogen atoms from ethane and replacing them with chlorine atoms (abbreviated Cl), a compound called *chloroethene*, or *vinyl chloride monomer*, is produced. This polymerises to produce *polyvinyl-chloride*, or PVC.

The refining of hydrocarbons also produces a lot of spare hydrogen – for example when hydrocarbons are fed into a cracking plant, or are used to

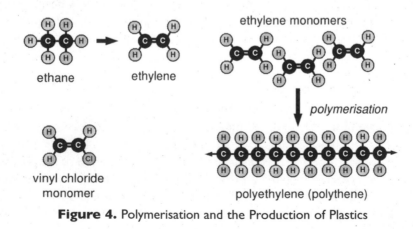

Figure 4. Polymerisation and the Production of Plastics

produce chlorocarbons. This hydrogen can be used in other processes. One significant process is the production of ammonia. By taking nitrogen from the atmosphere, one atom of nitrogen (abbreviated N) can be reacted with three atoms of hydrogen (H) to produce ammonia, NH_3. Ammonia is another important compound in the chemical industry, but one of the largest uses of ammonia is in the production of agricultural fertilisers. Without the use of hydrogen from the petrochemical industry energy would have to be used to produce hydrogen from other sources – an issue central to the idea of the hydrogen economy that is discussed in chapter 7.

So far we've just looked at oil, but the other important part of the petroleum industry is the production of natural gas. Natural gas is often held above the oil bearing structures of oil fields. This *associated gas* was once flared because it had no economic value, but today large amounts are collected and fed to gas distribution systems. Some geological structures contain little or no oil, but lots of gas, and are thought to be the result of hydrocarbon production from freshwater rather than marine organic matter. This *non-associated gas* is now used as the major source of gas for the natural gas distribution systems of industrialised countries. A good example are the gas fields of northern Russia and Central Asia, where natural gas is being extracted and transported by pipeline across the whole of Europe. Natural gas, made primarily of methane and ethane, can be used as a feedstock for the chemical industry. As oil runs short, more natural gas may have to be diverted to the chemical industry in order to produce hydrogen for use in vehicles, and to supply the carbon for the creation of carbon-based compounds.

One significant drawback of natural gas is that it's difficult to transport. It actually takes a lot of energy and organisation to move gas along pipelines, and even then the oceans are still a barrier to movement. For this reason one of the fasted developing areas in the gas industry today is the development of *liquefied natural gas*, or LNG, systems. LNG is produced by rapidly cooling natural gas to −162°C (−259°F) so that it turns into a liquid. However, the creation of super-cooled LNG requires a large amount of energy to generate a sufficient cooling load. The conversion efficiency of natural gas to LNG is about 85% [ECSSR, 2003] (that is, 15% of the gas fed into the plant is used to fuel the generators which super-cool the natural gas).

LNG is transported in specially designed, doubled hulled supertankers. These supertankers are not double hulled for safety, but in order that the LNG tanks can be insulated from the far hotter sea water. LNG terminals have been used in the USA, and in some European states such as France, Spain and Italy, for some time. However, it is only recently that the UK has begun to develop LNG systems on a large scale [Earth Cymru, 2004]. Here the gas is off-loaded and stored in insulated tanks. As the LNG takes up only one-sixth-hundredth of the volume of ordinary natural gas, LNG terminals can hold many days worth of gas supply in a liquid form. This also makes LNG an attractive proposition from the point of view of security of supply because LNG terminals can provide a large strategic reserve of natural gas in the event that other gas supplies are interrupted.

The other major use of natural gas since the late 1980s, apart from domestic and industrial uses, has been electricity production. Natural gas, like oil, contains a lot of energy. Unlike oil, and especially coal, it's also very clean to burn. Across the world more and more natural gas is being burnt in gas turbines to produce electricity. Given the versatility of natural gas some argue that this is a waste of resources. The other significant impact of the "dash for gas", in the UK at least, has been that it has enabled countries to meet their commitments under the Kyoto Protocol on climate change. This is because the increased efficiency of burning gas means you get more energy per unit of carbon released, reducing the level of carbon emissions from power generation systems.

From this brief review it's possible to see the depths to which petroleum has insinuated its way into our lives. Apart from the petrol and the gas, there's chemicals, paints, plastics, resins, man-made fibres – oil is everywhere around us. Remove it and a large part of our individual existence would disappear. It is the depths which oil has reached into our lives that leads on to another issue. Today oil economics dominates the globe. Most of the blue chip stocks on the stock markets around the world are directly involved with petroleum. Or, like the car, consumer electronics, chemical or pharmaceutical industries, their well-being is directly dependent upon the cheap availability of petroleum products.

To answer the question posed by the title of this chapter, *what is it about oil*, the answer has to be that for the past century it's been a mineral resource that has consistently produced a high return upon investment. It's become essential to the maintenance of the technological world, both in terms of the materials it provides, and the finance it generates. The trend in production of the petroleum market over the past century has consistently increased. Therefore any investment was guaranteed to make a return. Likewise, for governments, oil has been a consistent source of revenues either through the taxation of its use, or from the licensing of exploration and extraction rights. So, over the past one hundred years the whole of society, in the guise of the largest corporations and governments, has invested all its wealth in petroleum. The other critical factor is that the wealth produced has been invested in developments that require or support the use of oil. In many ways, the modern world has become addicted to petroleum and the products of petroleum [Observer, 2004a].

The other thing about oil is that *it can't last.* Although the reserves may last until 2050, from 2010 to 2015 it is likely that a growing shortage of oil production will cause prices to rise, and from that point on the current structure of the global economy will spiral into recession. As will be noted at various points in this book – it's the *First Law of Thermodynamics.* The energy of oil revenues, like the energy of the oil, has inflated the global economy. Take that energy away, and the economic system that oil supports must collapse.

Box 2. The First Law of Thermodynamics

The laws of thermodynamics describe how energy works within the natural world. The First Law basically states that the energy state of any system is dependent upon the amount of energy within the system. Therefore if the energy available for use within a closed system, such as the Earth, falls, then the energy-related processes within that system must become less energetic. In relation to global energy use, and in particular the use of fossil fuels, the drop in energy supply must create a consequential drop in the economic activity that fossil fuels enable.

---

2. Where Does the Energy Go?

Humans use a stupendous amount of energy, much of it sourced from fossil fuels. Each year, whilst economic activity grows, that level of energy use grows a few percent. In recent years one of the largest influences on the pattern of energy use has been the industrialisation of China and India. Where does all this energy come from, and where does it go?

In reality, it doesn't come or go. Throughout the universe energy and matter exist at a constant level – as explained in the *Law on the Conservation of Matter and Energy*. The fact that energy is constant is really useful when we want to examine how energy is used. In any system, if we measure the amount of energy that enters the system we know that the same quantity of energy must emerge somewhere. This concept of energy balance, that what goes in must come out, allows everything from a domestic boiler to the Earth to be studied and the movement of energy within that system mapped.

The use or storage of energy is dictated by the four fundamental forces that govern how the universe operates – the *strong nuclear* force, the *weak nuclear* force, the *gravitational* force, and the *electromagnetic* force. For this reason we can understand, and more importantly measure, the movement and flow of energy using physical laws.

The strong nuclear force provides the energy extracted from uranium in nuclear reactors. These nuclear processes create a lot of energy, for the use of very little fuel, because the strong nuclear force is the strongest force in the natural world. The weak nuclear force is a result of nuclear processes that

---

Box 3. The Law on the Conservation of Matter and Energy

The Law on the Conservation of Matter and Energy states that *energy in the universe is constant – it can neither be created or destroyed*. So energy can't be suddenly magicked into being. We must take one form of energy and convert or transform it into another.

Any technology we use to produce energy must, in turn, have a source of energy to draw upon, either from the environment or from the use of a fuel. This produces usable energy, and by-products such as heat, pollution and waste materials. In reality, what industrial society does is to take energy in one form and process it to produce a vary narrow range of other energy forms that we use as part of our everyday lives – for example motion (kinetic energy), gas pressures, heat, light, or electrical power.

---

take place when atoms decay. The Earth's core contains a lot of radioactive elements that decay to produce heat, and this *geothermal heat* causes the movement of the plates that form the crust of the Earth. In turn these crustal movements create volcanoes and hot springs. Like nuclear power, geothermal heat can also be used as a source of energy if the rocks below ground level are hot enough (usually in excess of 180°C to 200°C) to make heat extraction viable.

The gravitational force keeps objects on the ground, or makes them fall. Anything that has the potential to fall has *gravitational potential energy*. By utilising falling objects, like water in a water wheel, it is possible to extract the gravitational potential energy as usable energy. One of the largest sources of gravitational energy on the Earth is *tidal power*. As the Moon goes around the Earth it drags the water in the oceans across the surface of the planet, creating the tides. By trapping water in tidal estuaries, or by putting water turbines in the sea where tidal currents run fast, energy can be extracted.

The electromagnetic force is more complex, and works at many levels to provide sources of energy. The electromagnetic force operates through the electrons that orbit the atoms that make up everyday materials. It is the force that holds these materials together. It is also the basis of chemical reactions, and provides (or requires) energy when chemical compounds are reacted together. The electromagnetic force enables the flow of electricity. Electric current flows because electrons are transmitted between the atoms that make up conductive materials. The electromagnetic force can also be radiated as a beam of oscillating energy to produce X-rays, radio waves or visible light. The electromagnetic force also creates magnetism. This force is created when the crystals of magnetic elements, like iron, line up. This force can be created by man in the form of a magnet, or by using a coil of wire and electricity to form an electromagnet. It is also created naturally in the molten iron core of the Earth, creating the Earth's magnetic field. Most of the electricity we use is created in steam, gas, wind or water turbines that use spinning coils of wire in magnetic fields to induce the flow of electric current.

Possibly the most important type of energy to humans is *heat*. Heat is a property of matter created by the electromagnetic force. When matter is at absolute zero, −273°C, the atoms that make up the material are completely stationary − it is a *solid*. The electromagnetic force binds the atoms rigidly together, so they don't move. As heat energy is transferred to the material the atoms take on this energy as *kinetic energy* − they move around. Rather like springs, the heat energy stretches the electromagnetic bonds between the atoms causing them to oscillate to and fro. The hotter the material, the proportionately higher the level of heat energy that the material holds, and the higher the kinetic energy of each atom. Eventually, like over-winding a

---

Box 4. The Second Law of Thermodynamics

The Second Law of Thermodynamics states that *energy will only flow from a higher state to a lower state.* Think of energy like a river. The greater the drop in height, the faster the river will flow. In the real world this difference in energy states is created when we have a difference in temperature, or gas pressure, or voltages in an electrical circuit.

A consequence of the Second Law is that energy can't flow backwards – you can't flick a switch and make a river flow backwards up the hill. However, you could pump water back up the hill, but doing so uses more energy than was produced when the water fell down the hill. Even though making energy flow backwards is in practice a waste of energy, it does have some practical applications – for example the *heat pump*, described later in the book.

---

clockwork mechanism, the increasing kinetic energy causes some of the electromagnetic bonds between the atoms to break. At this point the solid material melts to become a *liquid.* Add yet more heat energy and more bonds break, and the material boils to become a *gas.* Finally, and requiring very large amounts of heat energy, all the bonds will break and even the electrons themselves will be dislodged, leaving the individual atomic nuclei moving randomly at high speed in a *plasma.*

How energy moves around our environment, transforming from one state into another, is subject to physical laws. Earlier in the book we met with the *First Law of Thermodynamics* – in short, the energy available in any system is determined by its inputs, and any change in input must change the energy available in the entire system. We need to add to this the *Second Law of Thermodynamics.* What this dictates is that to use energy we must create a difference in energy levels in order to generate a flow. Consequently how we can transform energy from one form to another is not a matter of guesswork. Energy transforms from one form into another proportionately to the way the transformation creates an imbalance in energy levels. So, if we know the magnitude of the forces entering the transformation process, we can calculate the forces that will come out. It's this ability to calculate, and therefore predict, the behaviour of physical systems that allows us to design energy production and transmission systems, and to calculate how society as a whole uses energy.

Some areas of environmental science are difficult to study because there is a chronic shortage of information to work with. For example, if you want to study the levels of contamination in processed food there's very little public data, and what data governments do collect they tend to keep secret. That's

---

Box 5. What is a Joule, What is a Watt?

Energy can be measured in various ways, but they are all based upon the international standard unit of energy, the *Joule* – abbreviated, J. A Joule is usually used as a measure of stored or potential energy, or as a measure of the heat energy contained in a material.

The Joule is a very small amount of energy. With reference to the Earth's gravity, it you take an apple weighing 0.1 kilos and lift it upwards one metre you have given the apple about 1 Joule, or 1J, of extra gravitational potential energy. Therefore energy measurements are often scaled by multiples of 1,000. To calculate the total value in Joules you take the figure given and multiply it by the value of the scale. For example, 2 kilo-joules, or 2kJ, the "kilo" part means 1,000 – so the energy measured is [2 x 1,000 =] 2,000 Joules.

Much of this chapter uses the exa-joule – that's 1,000,000,000,000,000,000J. Other parts of the book use kilo-, mega-, giga-, tera- and peta-Joules interchangeably.

Symbol	Name	Power	Value
k	kilo-	10^3	1,000
M	mega-	10^6	1,000,000
G	giga-	10^9	1,000,000,000
T	tera-	10^{12}	1,000,000,000,000
P	peta-	10^{15}	1,000,000,000,000,000
E	exa-	10^{18}	1,000,000,000,000,000,000

A Joule is a discrete amount of energy, and so it is usually used to describe stored energy, or heat energy. If we want to describe a flow of energy we have to scale the level of energy in Joules against time. Flows of energy are usually measured in *Watts*. One Joule flowing every second is equivalent to one Watt. Like Joules, Watts are small amounts of energy, and for this reason they are abbreviated using the same scheme described above – kilo-Watts, giga-Watts, etc.

Another scale often used in energy studies is the *Watt-hour*. This is equivalent to one Watt flowing for a period of one hour. As there are 3,600 seconds in one hour, and a Watt is a Joule per second, this means that there are 3,600 Joules in one Watt-hour. Like Joules and Watts, Watt-hours are also abbreviated, and this is how you get one of the most widely used measures of energy – the kilo-Watt-hour, or kWh (equivalent to 3,600,000 Joules per hour).

Another common measure of energy is the *British Thermal Unit*, or BTU – no longer used officially by the British but widely used in the USA. One BTU is equivalent to 1,055 Joules, and the *Therm*, made up of 100,000 BTUs, is equivalent to 105.5MJ (mega-Joules). The other common measure of energy used in energy studies is the *tonne* or *barrel of oil equivalent* – these are discussed in Box 8.

not the case with energy studies. Energy is big business, and so there's money to be made by producing information. Today there are a large volume of energy statistics available for free via the Internet. In fact, most of the sources quoted in this book are available online, and the links for each reference are provided in the sources chapter at the end of the book. Using these various information sources it's possible for the public to study independently the production and use of energy.

Most governments produce detailed studies on energy for their country. In addition, there are international agencies that specialise in collecting energy statistics at the regional and global level. For example, the International Energy Agency (IEA) and the Organisation for Economic Co-operation and Development (OECD) produce a variety of technical and statistical data on the current and projected future use of energy. Some companies, such as energy producers or financial analysts, also produce their own data. For example, British Petroleum (BP) produce a regular review of global energy statistics [BP, 2004].

To summarise the rather technical physics above, energy must have a source. If that source isn't immediately available and flowing, such as sunlight or the wind, then it must be stored. The most likely form of storage is as chemical energy, held in biomass (plant material), or fossil biomass like coal and oil. So, for any type of energy human society uses, we can ask a simple question... *where does it come from?*

The Sun is the Earth's principal energy input that, through the history of our planet, has enabled it to develop from a lump of rock to a place full of life. However, the amount of energy from fossil fuels the planet uses every year is over one quarter of the energy that the entire Earth receives from the Sun by photosynthesis every year – and therein lies the problem. If the energy supply is running out, whilst our current level of energy use is very large in terms of the planet's natural systems, how do you replace it from renewable energy sources?

The Earth is a *closed system* with a fixed set of energy inputs, so we can analyse the total energy that is available to the planet each year. Most of the Earth's energy comes from the Sun. The Sun gets its energy from the strong nuclear force, fusing hydrogen atoms to produce helium and other elements. This produces various forms of electromagnetic radiation that irradiate the solar system. Space is a vacuum so it can't transmit energy as heat, only as electromagnetic radiation (such as light) or as particles of matter that stream out from the Sun (the solar wind). The level of solar radiation delivered to the top of the Earth's atmosphere each year is about 5.4 million Exa-Joules (EJ) [OU, 2003] – or in plain numbers, 5,400,000,000,000,000,000,000,000 Joules. Of this total:

- 2.5 million EJ per year (46%) enters the atmosphere, heating the land and the oceans, in turn creating the *winds and ocean currents* (the winds and currents take-up 11,700 EJ per year) as well as powering the Earth's ecosystems via *photosynthesis*;
- 1.6 million EJ per year (30%) is reflected back into space by the atmosphere and the Earth's surface; and
- 1.3 million EJ per year (24%) runs the *hydrological cycle*, evaporating water to create clouds, rain and rivers.

The Sun's energy can be intercepted by natural systems in various ways. The land and oceans intercept the energy in sunlight using photosynthesis. This produces about 1,260EJ/year within biomass (plant material). In turn, this biomass forms the base of the food chain that supports the animal life of the Earth, including man. The oceans and atmosphere intercept the short-wavelength infra-red part of sunlight and hold it as heat. As the heated air creates differences in atmospheric air pressure, and water density in the oceans, this creates a flow of kinetic energy as the wind and ocean currents. Then, from a combination of wind and currents, waves. Of these, wind has been used by man for centuries, as both a static energy source (using windmills) and to power ships. Wave power has only recently been used as a source of energy.

The solar energy that evaporates water, creating the hydrological cycle, lifts water from the oceans to the tops of mountain ranges. In the process this gives the water gravitational potential energy. As the water falls back to

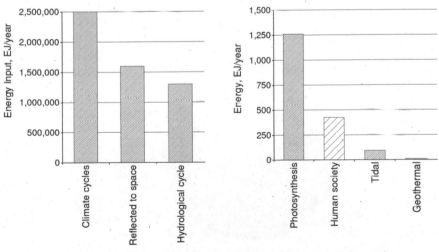

Figure 5. Natural Energy Sources (and human energy use)
[Source: OU, 2003]

the sea, it can be intercepted by water turbines, creating mechanical power or electricity.

In addition to the Sun's energy, *geothermal energy* contributes about 9.6EJ per year to the Earth's energy input. In some countries where hot rocks are close to the surface, such as Iceland, the Philippines, or the western states of the USA, this heat is utilised as a source of energy. The other significant input is the gravitational force of the Moon.

The Moon's gravity pulls on the water in the oceans, dragging them across the surface of the Earth. This lifts them against the Earth's gravity to give the tides their gravitational potential energy. This creates an input of kinetic energy of 93.6EJ per year. The scale of energy contained in the tides means that tidal currents or tidal streams are a significant source of energy – one that we have only recently begun to exploit.

To put the figures for the Earth's solar input into perspective, the United Nations Development Programme (UNDP) estimates that the global population uses 424EJ of energy per year [UNDP, 2000]. That's 0.01% of the energy that reaches the Earth's atmosphere, but around one-third of the energy produced every year by photosynthesis. Of this 424EJ, 34.6% is sourced from petroleum, 21.6% from coal, 21.4% from natural gas, 11.3% from traditional biomass (firewood), 6.6% from nuclear, 2.3% from large hydroelectric schemes, and 2.1% from other renewable sources. Therefore 77.6% of the world's energy (329EJ/year) is sourced from fossil fuels. With current trends in energy use, 56% of human energy use (the 237EJ/year of oil and gas) is

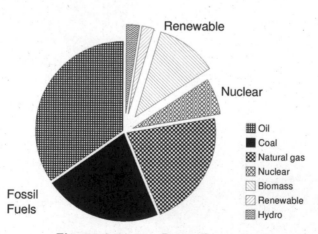

Figure 6. Human Energy Sources
[Source: UNDP, 2000]

going to run out or be scarce within the next one hundred years, and one third of it (the 147EJ/year of oil) will become very expensive within the next fifteen to twenty years. It would be possible to switch to coal for a large part of this energy, but that cannot happen because of the impacts on climate change.

Energy economics, the study of the way energy is sourced, processed and utilised by society, uses different methods to express energy production and use. As energy economics utilises complex information, rather like other forms of statistical process, the interpretation of energy studies can be difficult. In order to make the correct decisions about our energy future it's important that as a society we can have confidence in the information that is given to us. Therefore, rather than understand the detail of these studies, what is more important is to understand the process by which these studies are carried out, and through this understanding, be able to ask challenging questions of those producing the information.

Most energy studies divide the flow of energy according to the input and output, or the supply and demand or consumption. Large scale studies may also define whether the energy source was in its raw *primary* form, or a processed *secondary* form (see Box 6). Primary energy is assessed by adding together the energy value of the raw fuel supplied, such as coal or gas. Secondary energy is things like electricity, which is produced from primary energy sources, but at a lower level due to the inefficiencies of converting one source of energy into another. However, this split between primary and secondary energy can create confusion when we look at renewable sources. For example, wind turbines and hydroelectric plants don't have a fuel input. And whilst nuclear power stations use nuclear fuel, it doesn't really assay in the same way as the chemical energy of oil. Therefore although they produce electricity, the electricity produced by nuclear or renewable energy technologies is classed as a primary form of energy.

When looking at countries or large regions, energy studies will also consider energy imports and exports, and perhaps conversion or distribution losses, in order to give a better view of the total energy flow. How imports and exports affect the overall balance for the system depends upon how well the organisations who are importing, processing, and then perhaps re-exporting energy commodities report their data. If there is leakage in the system, for example in states where oil or coal are imported or exported on the black market, then this will invalidate the results of the energy assessment. Likewise problems with the assessment of conversion losses, since this is nearly always statistically significant, can also skew the results of energy models.

There are various models and statistical digests of energy around. The most easy to understand model is BP's *Statistical Digest of World Energy* [BP, 2004].

Box 6. How Energy Models Define Energy Use

In the energy studies produced by governments and international agencies a standard model is used to define how energy is sourced and utilised. This defines energy as either a primary or a secondary source. On larger models other factors, such as the effects of imports and exports, may also be considered.

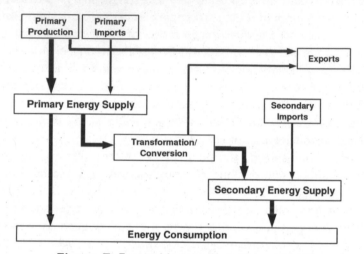

Figure 7. Energy Movements in Energy Models

The primary energy supply is made up of all the energy sources provided to the model, not including the value of energy exports. This will include oil, gas, and coal. Primary electricity, mostly generated by nuclear and renewable energy sources, is also included in this figure. However the supply of these energy sources will usually involve a small percentage loss due to losses within the system – for example gas leaks or oil spills. There are also various conventions applied on how the primary energy produced from nuclear and renewable sources are included in energy models, and sometimes this will create an inconsistency between how much energy was produced from these sources, and how much was in reality contributed to the energy economy.

A proportion of the primary resources will be converted or transformed to produce the secondary energy supply. The level of the secondary supply will depend upon the efficiency of energy conversion. Finally, the sum remaining primary supply and the secondary supply is conveyed to energy users and consumed.

To interpret the information produced by the UK Department of Trade and Industry and others it helps if you are able to understand how the definitions of *primary* and *secondary energy* are applied.

This provides information on different forms of energy use in different states around the world, and because the statistical method is consistent between states you can compare energy use in one country directly with another.

Looking at human energy consumption regionally there are large variations. For example, the average person on the planet uses 58.6 giga-Joules (GJ) of energy per year. The average resident of North America uses 272GJ/year, in the Europe and the former Soviet Union 130GJ/year, and in the "Rest of the World" (that is not part of North America, Europe or former Soviet Union) 33.5GJ/year. So, a resident of North America uses over eight times the energy of a resident of the Rest of the World, or more than twice the energy of a resident of Europe. The use of energy in the industrialised world isn't just represented by driving cars, or heating (or cooling) homes. A large part of it is made up of lifestyle-related uses of energy, such as the use of processed food or the use and disposal of packaging materials and other single-use goods. So reducing the energy supply, or raising the cost of use, will hit lifestyles, not just the ability to drive or fly.

It might be easy to believe that the effects of Peak Oil would affect the industrialised world more than the developing world – because the industrialised world's higher use of energy would leave them more exposed to the higher price of oil. This is not the case. Yes, in the longer term, those in the industrialised world will experience the greatest change in their living standards, but in the short term it will mostly be developing countries that will feel the effects. There are a number of reasons for this. For example, crude oil is usually priced in the currency of the industrialised countries, such as the Euro or the US Dollar, and the currencies of developing countries are weak against these major currencies. As a lot of the oil and oil products they use are also refined in the industrialised states, this increases the cost of energy in developing states. In addition to the total level of energy use what's also important is the fuel mix – the balance of oil use to other fuels such as coal or gas. Those countries that are not heavily reliant on petroleum for energy will have a far better time in the coming post-Peak Oil crisis. The fact is that, despite their low level of industrialisation, developing countries are far more reliant on petroleum in terms of their total energy use.

From the figures in the BP dataset [BP, 2004], if we divide the level of petroleum consumption as part of primary energy supply by the total primary energy supply for that country, then the ratio produced represents the sensitivity of that country's economy to the price or availability of crude oil. If we look at the results at the regional level, the most affected region (the region with the highest ratio) is the Middle East (0.50 – half their energy comes from oil). Energy is crucial for them not just for transport, but because it fires the desalination plants that provide freshwater. The Middle East is followed by

Table 1. Ratio of Petroleum Consumption to Total Energy Consumption
[Source: Adapted from BP, 2004]

Top 10	Bottom 10	Other major states
Singapore (0.88)	Bulgaria (0.22)	15. Spain (0.53)
Ecuador (0.78)	Turkmenistan (0.22)	18. Italy (0.51)
Other Middle East (0.68)	South Africa (0.21)	20. Netherlands (0.49)
Kuwait (0.65)	Czech Republic (0.20)	24. Denmark (0.47)
Other S. American (0.63)	Slovakia (0.19)	33. USA (0.40)
Hong Kong (0.62)	Kazakhstan (0.19)	37. Germany (0.38)
Portugal (0.62)	Russian (0.19)	41. France (0.36)
Philippines (0.61)	Qatar (0.14)	43. UK (0.34)
Greece (0.61)	Uzbekistan (0.13)	49. India (0.33)
Republic of Ireland (0.60)	Ukraine (0.10)	59. China (0.23)

South and Central America (0.47), then Africa and North America (0.40 each), then the Asia-Pacific region (0.36), and finally Eurasia (Europe plus Russia/Central Asia, 0.32 – only 32% of their energy comes from oil).

Of the seventy countries or groups of countries used in the BP survey, if we rank the ratios from the highest to the lowest then it is possible to see how different states will be affected (see Table 1). Certain states, such as Singapore, will be deeply affected by changes to the oil price because they are so highly reliant on petroleum. Others, because of their low level of development, or because their economy is based on other sources such as coal, will be less affected.

Europe uses a lot of energy compared to the rest of the world's population. Western Europe as a whole consumes 16% of the world's energy supply whilst having 15% of the world's population. In contrast Africa consumes 5% of the world's energy supply whilst having 13% of the world's population.

As well as variations between regions around the globe, there is a wide variation in energy use within Europe. Rather than considering energy consumption, figure 8 shows the *total primary energy supply per person* (TPES) for various European states.

When considering the effects of energy shortages on a country, or the need to reduce fossil fuel use because of climate change, the primary energy supply is a more suitable measure. This is because it describes the *energy intensity* of the whole economy. This will include the manufacture of goods for

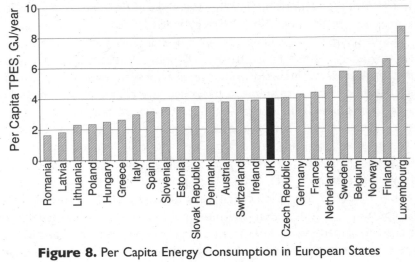

Figure 8. Per Capita Energy Consumption in European States
[Source: IEA, 2003]

export, not just the domestic energy consumption of its population. In general there is a two-way division of energy use in Europe. Countries in northern Europe use more energy than those in southern Europe due to the extra demand for Winter heating. At the same time the more developed economies of western Europe use more energy than those of eastern Europe. Even with these two general trends there are still big differences in per capita energy use between European states because of the different levels of development or environmental performance in each state.

The UK falls within the mid-range of European per capita primary energy consumption. Even so, the scale of the UK's energy use is still large. If we use the BP dataset [BP, 2004] to provide a consistent comparison, then global primary energy consumption in 2003 was 408EJ/year, of which the UK's share is 9.2EJ/year. This means that the UK's 1% of the global population uses 2.3% of the world's primary energy – or, on BP's statistics, over twice the energy used by the whole of Africa. In fact, the energy the UK wastes from its coal- and gas-fired power stations, around 2.24EJ/year, is about the same as the whole primary energy consumption of Belgium and Luxembourg, 2.8EJ/year (albeit Belgium has one sixth the population of the UK), and larger than Egypt's primary energy use, 2.18EJ/year (even though Egypt has a slightly larger population).

The data provided by organisations such as the International Energy Agency, or BP, are intended to provide a global overview. When looking at the UK's use of energy these do not provide a lot of detail. For that you have to look at the information produced by the UK's *Department of Trade and Industry*

(DTI). The DTI are responsible for the oversight of the UK's energy industry. They produce regular bulletins of energy statistics for the UK. By analysing this data it's possible to evaluate the UK's use of energy in more detail.

The way energy is presented to the UK public by the media — whether it is wind turbines, waste incinerators, or nuclear power stations — is as domestic electrical power. For example, a particular plant produces X *many households-worth* of power. This is a fallacy, in part because domestic energy consumption is the smallest part of the UK economy, but also because electricity isn't the most widely used fuel. In fact, electricity only forms a minor share of the UK's energy usage in terms of *primary energy* sources. Just 9% of the UK's primary energy supply [DTI, 2004a] was primary electricity (mainly nuclear power) in 2003, and overall electricity only made up 18% of the UK's energy consumption [DTI, 2004b]. For the most part electricity is a *secondary* form of energy produced inefficiently from primary energy sources, and the total energy of the fuels transformed into electricity are less than the gas, coal and oil used directly by industry, in homes, and in the transport sector.

With the coming difficulties in energy supply what really matters is the *fuel mix*. When discussing energy sources the point often made by government, or those promoting unpopular sources of energy like nuclear power, is that the more diverse the UK's energy sources are the more secure its energy supply is. However, despite the recent claims of government, the diversity of energy sources in the UK is dwindling as natural gas becomes the main source of energy for homes, industry and for electricity generation. These trends can be seen if we look at the annual change in primary energy sources in the UK. Figure 9 shows how gas has slowly gained in the share of UK energy sources at the expense of coal. It also provides a comparison between the energy provided from fossil fuel and renewable sources (renewables are the thin band at the top of the graph that begins to show up around 1987).

In 2003, 90% of the UKs primary energy supply was sourced from fossil fuels. 41% came from natural gas, but only 32% came from petroleum because in 1996 natural gas eclipsed oil as the largest source of fossil fuel energy in the UK economy. Petroleum is mainly used in the transport sector, not for energy production. For this reason there has only been a small change in the level of petroleum use over the past 30 years, mostly as a result of industry switching to gas (from gas oil or kerosene) for raising heat. Coal, whilst still making up 17% of the primary energy supply, is shrinking as a source of energy because coal-fired power stations are being closed in favour of more efficient gas-fired systems. Nuclear power, has reached its zenith. Over the next twenty years most of the UK's nuclear power plants will close, and there is no official policy for their replacement. Despite the fact that nuclear represented around 22.4% of electricity generation in 2003, it only represented

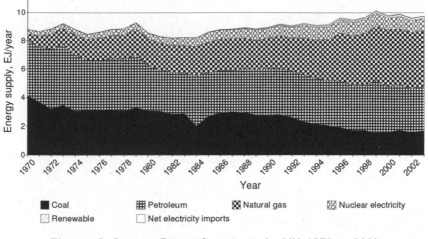

Figure 9. Primary Energy Sources in the UK, 1970 to 2002
[Source: Adapted from DTI, 2004c]

8.6% of primary energy supply [DTI, 2004a] because electricity use only forms a minor part of overall energy use in the UK.

If we look at how energy is utilised in the economy there is a crucial difference between the primary supply and the final use of energy. This is seldom communicated to the public as part of the debate about energy use, but it is crucial to how the UK uses energy. Figure 10 shows the levels of *primary energy supply* and the *final use of energy by fuel* in the UK during 2003. By placing a column representing primary supply next to final use it's possible to see how energy transfers from supply to use. Each column represents the same amount of energy, and in accordance with the *Law on the Conservation of Matter and Energy* what goes in must come out. This makes it easy to see another factor that's not reflected in discussions of energy policy that concentrate on final supply – *system losses*.

Between the primary supply of energy and the final use of energy the gas and coal diminish because their energy is being transformed into electricity – a large part of this energy is shifted into the 'system losses' segment because it is lost as heat from power generation. If we ignore system losses, then electricity represents 18% of final use, gas 36% and oil 42%. However, if we include system losses, then, of the original energy supplied to the UK economy, electricity represents just 10%, gas 26% and oil 30% – 28% of the energy is lost from the system, mostly from electricity generating plants.

Note that in figure 10 the oil segment changes very little – almost all goes toward final use. This is because most of the oil supply goes straight into the

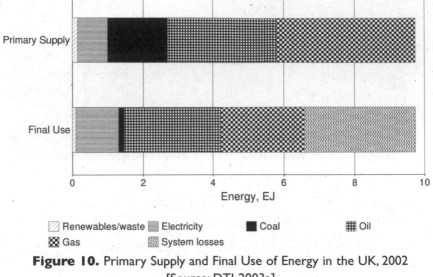

Figure 10. Primary Supply and Final Use of Energy in the UK, 2002
[Source: DTI 2003a]

transport sector.

When examining how our use of energy is accounted for it's important to look at the whole picture. That means examining where the energy used in the UK comes from and how it flows through the economy, not just where it ends up.

Any energy policy, such as the recent expansion of renewable energy sources, that concentrates primarily on producing electricity will only solve the energy needs at the beginning of the supply chain – usually called the *supply-side*. Instead energy policy needs to focus on the whole system of energy use, and seek to reduce the high levels of energy demand by consumers – usually called the *demand-side*. That's not a politically popular solution since it requires rethinking a lot of the methods by which society uses energy.

As well as tabulated data, the DTI also produce flow diagrams that illustrate how Britain uses energy [DTI, 2002a]. If we simplify this information and update it with the most recent data [DTI, 2004b], as illustrated in figure 11, it is possible to see where the UK's energy is sourced, and where it is used.

Figure 11 is a complex diagram. What it shows is energy entering the UK economy on the left, and flowing to the consumers on the right. Losses from the system leave at the bottom. However, the various flow lines are all drawn to scale, so that the quantity of flow is proportional to the thickness of the line. This means that the diagram gives a direct visual representation of energy flows through the UK economy.

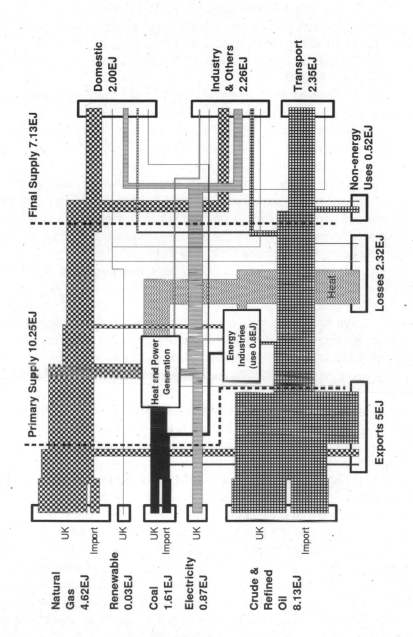

Figure 11. A Flow Diagram of Energy Use in the UK, 2003
[Source: Adapted from DTI, 2002a/DTI 2004b]

How you assess the UK's primary energy supply depends where you draw the dashed line on the left hand side. The DTI do not include the value of energy exports – both primary energy exports like crude oil, as well as secondary energy such as refined fuels. Therefore while the total value of the energy that flowed through the UK in 2003 was just over 15EJ, the primary supply was about 10.25EJ.

The source boxes on the left of the diagram split the energy entering the system according to fuel type. They also denote whether that energy has been produced from within the UK (UK), or whether it has been imported (Import). The electricity source is mainly nuclear power, and a small import of power via the *Channel Interconnector* – which allows the import of electricity from France. There is also a small input from renewable sources (0.11EJ). However most of this is made up from waste fuels destined to be burnt and so only one quarter, 0.03EJ (the figure used in the flow diagram) of this is actually consumed as part of the final energy use.

Once the energy has entered the economy it flows to the consumers. Petroleum products and gas flow direct to the consumer, but a proportion of the energy – almost all the coal, and nearly half the natural gas – is transformed to produce electrical energy. The transformation produces energy as electrical power, and a larger quantity is lost as heat in the cooling water and steam plumes of power stations (this is shown as the heat flow from the Heat and Power Generation box). As yet the UK does not widely use *combined heat and power* (CHP) systems. This allows the heat from energy transformation to be used as part of another process in order to decrease the amount of energy lost without being put to a productive use. There is a national CHP programme, but as most of this capacity is in the commercial sector, and because most of the plant is gas-fired, it is merely shifting dependence onto natural gas for a greater part of the UK's economic activity.

The energy distribution system as a whole also loses energy. Electricity is lost as heat from the power grid, gas leaks from the distribution system, and oil spills from pipelines and tankers. These system losses are the other flows that enter the losses box at the bottom of the diagram. Using this model, in 2003 roughly one quarter (2.32EJ/year, or 23%) of the UK's primary energy supply was wasted as system losses, mostly as heat from energy transformation.

Some industries, such as the iron and steel industry, consume large quantities of energy. These are collectively referred to as "Energy Industries", and consumed 0.8EJ of energy in 2003. However, although they contribute to the UK economy, their operation is kept outside the UK's final energy supply, even though the 'non-energy' uses of energy minerals (0.52EJ in 2003), like lubricating oils or industrial gases, are included within the final supply figure. About 3.12EJ is lost between the primary supply and the measure of final supply –

which in 2003 was 7.13EJ. Of this figure, the largest amount, made up almost entirely of petroleum products, goes into the transport sector (2.35EJ). The industrial (2.26EJ) and domestic (2.0EJ) sectors have a more diverse supply of energy, but the domestic sector is more dependent upon natural gas.

What the flow diagram represents is the movement of energy – as the energy content of oil, gas, coal, and the electricity from nuclear and renewable sources. But this isn't the entire picture. The UK imports raw materials such as steel and plastics, as well as components to make goods, and assembled consumer goods. There is no data available on the value of this *hidden energy*. The energy that these products contain is attributed to countries where these goods were manufactured or the raw materials made. This means that states like China, who have a large energy use (mostly sourced from coal) within manufacturing industry, and who are importing more energy in the form of crude oil, are not directly using that energy. It is being passed in the form of manufactured goods to the states where those goods are consumed. Hence, within the UK, not all the energy value of fuel used for manufacturing products, such as the gas or oil used in the manufacture of petrochemicals, plastics and fertilisers, is completely accounted for in energy statistics.

So far, for the sake of comparison, we've been looking at energy in exa-Joules (EJ). The DTI's figures are usually expressed in *million tonne oil equivalent* (mtoe) – 1mtoe being equivalent to 0.042EJ. Using the DTI's data, in 2003, the UK used an amount of energy equivalent to 248.1 million tonnes of oil [DTI, 2004b], or 10.4EJ – enough energy to keep a 100-Watt incandescent bulb lit for 3.3 billion years. Beyond this headline figure for primary energy use there are many other DTI studies that describe other trends within the UK energy economy [DTI, 2002b]:

- Domestic energy consumption has risen 32% since 1970, and 19% since 1990 (that is, energy use over the last 12 years of the study rose 50% more than in the first 20 years). The energy use in households has increased 19% since 1990, partly because of more appliances per household, but the higher number of households is probably more significant.
- Space heating and hot water accounted for 82% of domestic use, and 64% of commercial uses.
- Energy use in the industrial sector fell 55% between 1970 and 2001, primarily as a result of the shift away from energy intensive manufacturing industries towards the service sector.
- Energy consumption in the transport sector has almost doubled since 1970, and tripled in the air transport sector over the same period. Between 1990 and 2001, energy use in the air transport

sector increased 56%, rail transport 8%, and road transport 7%. Since 1990, the energy use in road freight has increased by 17%, whilst passenger transport only increased 1%.
- Between 1970 and 2001, energy consumption in the service sector overall rose 25% – but whilst in the public sector energy consumption fell 7%, in the private sector consumption rose 59%.

That's a large dollop of statistics but it can be summed up in a single word... *more*. As the UK economy grows we consistently use more energy. Two factors drive this: economic growth and technological development. Economic growth is a strong driver of energy demand. The more activity there is in the economy the more energy is used. Technology works in two directions. Often technology is portrayed as reducing the intensity of energy use by making machines more efficient, or reducing waste. This is true, but technology also makes manufacturing easier and cheaper, thereby allowing more units to be made, sold, and transported further. This increases demand for commodities, and consequently the net energy use.

Despite what may be said about the efficiency of new technology, when we look at technology we have to look at both dimensions of energy demand to discover the true effect. For example, the fuel efficiency of cars in the UK has improved over the last decade, saving the equivalent of 0.5mtoe (million tonnes of oil equivalent) per year [DTI, 2002b]. But at the same time the ownership and use of cars and the distance they travel has increased the equivalent energy use by 0.9mtoe per year. So the overall effect was an increase in car-based energy use of 0.4mtoe – nullifying the improvements in engine efficiency.

The mode of energy use is also important to overall efficiency. Often this requires that we rethink how we perform certain kinds of activity in society, not just how the activity itself operates. For example, of the energy burnt in a car engine only 25% is turned into motion along the road. With current technologies it would be difficult to create a car engine that is significantly more efficient than this. But if we view this use of energy in terms of its purpose, in this case the energy expended *per person moving* down the road, then we can half or even quarter the energy use (increasing the efficiency by a factor of two or four) by putting two or four people in the car instead of just one.

Another example is the way our consumption of food has changed over the last ten to fifteen years. It is more efficient, in energy terms, to heat food with a microwave oven than with a conventional oven. Of course, some people refuse to use microwaves because they they prefer food to be cooked rather than heated. However, if we assess the whole food system, including the energy used in food production, transport, preparation and use, people that cook using ovens at home probably use a lot less energy. This is because the types of pre-prepared foods cooked in a microwave often source their ingredients from across the globe, creating a large energy sink in the form of transport. The produced food may also have been transported long distances. More significantly, the food has already been cooked, and then refrigerated, and then kept refrigerated for long periods, before the consumer heats it up again in the microwave. By contrast people who cook their own food often buy raw ingredients, which are usually sourced closer to home than food produced on an industrial scale, and they cook and consume the food immediately – consequently they use less energy overall.

Such social factors are an important component in energy analysis, and will be an important component in finding ways for society to adapt to a lower level of energy use when oil and gas become scarce. When considering social factors we also have to consider people's preferences for how they use energy, and how people's lifestyles determine how much energy they use. At the same time economic factors, as a deliberate matter of public policy or due to market conditions, may also influence people's choices on energy use. Therefore when making an assessment of how certain activities affect energy use we have to look at the range of activity within the population, and look at the different energy impacts that different modes of activity create.

This is the problem the UK faces. Individual devices or systems may be more efficient, but as we produce and consume more those savings are swamped by the increased demand for energy. So, even if the message from the Department of the Environment is "save it", the message from the DTI is that the UK's energy use is projected to rise by an average 2% or 3% per year each year for the next decade [DTI, 2002c]. However, if we do face shortages of energy within the next ten to twenty years as we pass the date of Peak Oil, current trends will only make matters worse. We will become more dependent upon natural gas, forestalling perhaps a greater crisis when gas reaches its own projected peak in production between 2020 and 2030 [POST, 2004].

It's fairly obvious that in a world where energy resources are finite we can't continually increase our production forever. What's important is the level of production that can be sustained. At this point our currently *unconstrained* use of energy resources will be constrained by the physical restrictions on the production of that energy resource. In the case of oil that doesn't just mean that growth is reduced, it will actually lead to a sharp fall, year on year, in the oil available.

To understand the effects in more detail we need to project our use of energy into the future, based upon current trends in energy use. We also need to know more about the time-scale for the decline of oil, and so we need to examine why different dates for Peak Oil exist. At some point these two factors – consumption and production – will intersect in time, and this is the point at which our consumption will be limited by the level of oil production. What's also important, as around 90% of the UK's current energy supply is based upon fossil fuels, is to understand the debate over climate change and how this might affect our use of energy. These are the issues covered over the next two chapters.

3. Peak Oil – Why and When?

As noted earlier, given the physical restrictions on oil production, the key question is not when will oil run out, but when will our demand for oil exceed the capacity of the world's oil reserves to provide the oil we use. This factor is more important than the date at which oil will run out because when demand exceeds supply the world price of oil will rocket. What we need to discover is when will oil production peak, and what will the price of oil reach when demand exceeds supply.

In early 2004, oil prices began to rise across the globe. By May 2004, they had reached a record dollar price – although if you take into account the effects of economic inflation over the last twenty years the price was only just over half the value of the record prices of 1980 [BP, 2004]. These high prices, despite the best efforts of politicians and the oil market, were maintained throughout 2004 and into 2005.

Why these prices rises occurred were largely due to political uncertainty, in part driven by concern about the growing dependence upon Middle Eastern oil production [BBC, 2004a], and speculation on the future price of oil by financial markets [Guardian, 2004b]. For example, the situation in Iraq following the Second Gulf War, problems with unions in oil producing states like Nigeria and Venezuela, and terrorism in the Arabian Peninsular all affected the oil price during 2004. However, underlying the political factors is a more significant trend.

Oil consumption is rising [BBC, 2004b], but oil production, and the refinery capacity for certain oil products, is not keeping pace with this rise. Consequently the margin between the oil and oil products the planet produces, and what the planet demands, is diminishing. This means that political factors, such as terrorist attacks, might cause a short-term change in oil prices over a few days. However, these short-term changes are taking place on the back of a longer term rise in prices caused by a perceived threat to global oil supplies created by a lack of production and refining capacity.

In June 2004, and on a number of occasions during the rest of 2004, the Organisation of Oil Exporting Countries (OPEC) agreed to raise production quotas to meet the increased demand for oil. This caused prices to drop briefly, but what if such a rise in oil production were not possible? What if OPEC, which represents the states who sit on the world's largest reserves of conventional oil, were physically unable to turn up the taps? This is the question posed by *Peak Oil* – the point at which the physical constraints on conventional oil production limit the amount oil producing states are able to supply.

Box 7. Understanding Different Types of Oil Production

One problem with talking about Peak Oil is that those involved are not necessarily talking about the same types of oil. To understand the arguments, it's necessary to understand what people are talking about when they refer to *primary, secondary* or *tertiary* sources of oil.

The debate on Peak Oil relates to *primary oil* (also described as *conventional liquids* or *conventional oil*). This is liquid oil that is pumped out of underground reservoirs. This type of oil has a peak because of the geophysical properties of oil reservoirs.

Secondary oil (also called *non-conventional oil*) is made up of forms of oil that are locked-up within mineral deposits, and these must be mined like other minerals such as coal or iron. Production from these sources is limited only by the capacity of humans to dig these minerals out of the ground, and so there is no effective peak in their productive capacity. However, given that producing non-conventional sources requires large-scale mining operations over large areas of land, there will always be a theoretical limit to production dictated by the local geology and the economic viability of mining the resource.

Finally *tertiary oil* (also referred to as *natural gas liquids* or NGL), is produced from other production processes. Tertiary oil is mostly produced by the refining of natural gas. Some estimates of Peak Oil include the tertiary sources of oil, whilst some do not. As so much gas is processed today NGL is a significant contributor to oil production in countries such as the USA, but in reality NGL will only provide a few percent of the energy resources provided by conventional oil.

The price of any commodity is determined in inverse proportion to its availability. That is, the less there is, the more it costs, or vice-versa. So, when the oil producers can't physically pump more oil to satisfy the rising demand the price of oil must rise. This isn't just a matter of supply economics. It also has a lot to do with how the oil will be produced. As we work down the reservoir of global oil reserves, those that are left will cost more to produce and deliver to the consumer.

As described in chapter 1, oil comes in many forms. Most of the oil deposits that the world relies upon today are made up of oil held in sedimentary basins. These are described in assessments of oil reserves as *conventional liquids*. The pore spaces between the grains in the sedimentary rocks allow fluids, like water and oil, to move through them (see figure 12). In addition any fissures or voids running through the rock allow fluids to flow at higher speed than via pore flow. The combination of fissure flow and pore flow provides a physical speed limit to the movement of oil, depending upon the pressure

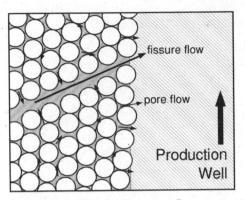

Figure 12. Oil Flow in a Reservoir

regime that acts upon the oil field.

As the oil does not mix with the water below the reservoir, the pressure of the water beneath the oil forces it to the surface when it is tapped with an oil well. In fact, one of the earliest measures of the size of an oil reservoir was the period of time that oil would gush from the well when the first well was sunk. The pressure is created by the buoyancy of the oil relative to the water saturated rocks around it – rather like the hole in the water a steel ship makes (called its displacement) which creates enough buoyancy to keep the steel ship afloat. The release of pressure when the oil is tapped also causes gases compressed within the oil to expand, which also assists its movement upwards towards the surface. You can see the same effect when you open a bottle of fizzy drink and it squirts out around the cap of the bottle.

Oil production plants also pump water into the rocks beneath the oil reservoir to increase the pressure level and make the oil move faster. Even so, oil will only flow at a certain rate due to the flow properties of the rock that it is held within. Once that rate is reached with the available means of extraction oil cannot be produced any faster. The combination of upward water pressure, the buoyancy of the oil, and gas pressure within the oil, causes the oil to rise, but once you take enough oil out of the reservoir the loss of pressure and buoyancy is also what causes the flow of oil to slow to a trickle. Even pumping oil from the reservoir, creating a negative pressure around the oil well, will make little difference. You can see why by trying to suck the air out of a glass bottle (you can only get so much air out of the bottle before it becomes impossible to suck any more). Oil reservoirs are similar because the time delay created by the speed at which water flows through the rock beneath the oil reservoir limits the movement of oil into the extraction wells.

A single oil reservoir will have more than one oil well sunk into it.

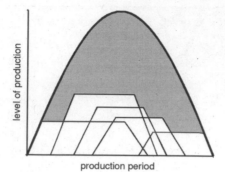

level of production

production period

Figure 13. Oil Production and Reservoir Lifetime

Following the initial pumping of the well, during which the production rate rises as the system of pore and fissure flow is established, oil production rises until it reaches the maximum production capable from the well (see figure 13). As pressure within the reservoir is localised, each oil well produces at a different rate. Together the collection of wells allow more oil to be extracted from the reservoir than if only one well were used, but when the reservoir reaches about half its productive capacity the pressure across the reservoir begins to fall. This affects production from all the wells, and sinking extra wells makes little difference as the pressure regime across the oil reservoir has been changed.

Gradually the production level from the whole reservoir falls until a point where further extraction is not economic. This is the situation now being experienced across the USA's oil fields [BBC, 2004c], and more recently in the British sector of the North Sea [Guardian, 2004c]. The reduction in output from oil reservoirs can't be avoided – it's a physical feature of the way the oil flows through the reservoir rock. This means that even though a reservoir may have a theoretical capacity based upon its volume, what can be produced is a lot less than this. It's likely that only 50% to 60% of that capacity can be extracted by conventional means, and in practice, the last half of the potential production capacity will take a lot longer to extract than the first half, reducing production volumes during the later life of all oil fields.

The peak in production from a particular oil field is usually accompanied by a change in ownership. Whilst some large companies specialise in exploration, and some in the first stage of production, the maintenance of a falling level of production from an oil well is only an attractive proposition to smaller companies prepared to accept the lower profit margins from oil fields that have past their peak.

Box 8. Oil Barrels, Tonnes and Potential Energy

The most commonly used unit in energy economics is the *tonne of oil equivalent* – abbreviated *toe*. This unit of measurement expresses all energy as the energy contained in one metric tonne of average refinery crude oil – averaged because different types of crude oil have a slightly different energy content.

1toe is equivalent to 41,868,000,000 Joules, or 41.868GJ (giga-joules). Expressed as a volume, one tonne of crude oil is equivalent to 262 imperial gallons or 1,192 litres. As society uses so much oil the tonne of oil equivalent is also expressed as the *thousand tonne of oil equivalent* (ttoe) or the *million tonne of oil equivalent* (mtoe). 1mtoe is equivalent to 41.868PJ (peta-joules), or 0.041868EJ (exa-Joules).

There is also the older measurement of the *barrel of oil equivalent* (boe). By volume, 1 barrel of oil equals 42 US gallons, or 34.97 imperial gallons, or 159 litres. By mass there are seven and a half barrels of oil to one tonne of oil. Expressed as energy, one barrel of average refinery crude oil equals 5,710,000,000 Joules, or 5.71GJ. Again, because the barrel is a small measure, it is usually used in multiples of millions or billions.

Currently (late 2004) the world is consuming around 83 million barrels of oil per day, which equates to around 30 billion barrels of oil per year.

The principles that govern the extraction of conventional oil were identified in the 1950s by the American geophysicist, M. King Hubbert. In 1956, Hubbert presented a paper to a conference of the American Petroleum Institute [Hubbert, 1956] in which he predicted the future production levels from American oil fields, taking account of the physical restrictions on production from the ageing oil reservoirs. To make this prediction he analysed the production records of many different US oil fields from the middle of the Nineteenth Century. In particular, how each oil reservoir performed over the lifetime of the oil field. He also identified from this study that American oil demand in the 1950s was exceeding the level of new discoveries, and therefore that the oil reserves discovered within the USA would not last as long as initially believed because of the increase in consumption.

Whilst dismissed at the time, Hubbert correctly predicted the peak in American oil production in 1970, and the subsequent fall in production. His projections, and the graph presented in his paper, have since been called Hubbert's Plot or Hubbert's Peak (see figure 14). The value of Hubbert's method isn't just the prediction of when production will peak. It also provides an analysis of the potential for new finds, based upon historical experience,

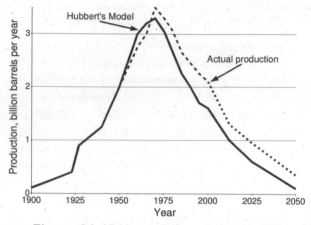

Figure 14. Hubbert's Plot or Hubbert's Peak

and the levels of production that might be achieved from the existing reserves and any new finds. Hubbert's 1956 paper assessed US oil reserves at 200 billion barrels. The actual figure is around 220 billion, but this extra 20 billion barrels is attributable to the production of natural gas liquids (NGL) which were not considered in Hubbert's original study.

Hubbert's plot was based upon conventional oil – free oil held in sub-surface reservoirs. However, there are other secondary sources of oil and gas, outlined in chapter 1, that can supply oil. These produce non-conventional liquids that are not so dependent upon the flow properties of oil reservoirs because they are physically mined. The problem with non-conventional liquids is that they cost far more to produce because the process is more labour intensive, and it uses a lot of energy. In relation to Peak Oil, and the world's maximum level of production, there is also the issue that these sources do not yield as much oil for the extra effort involved.

Non-conventional sources also have a higher environmental impact because of the effects of large-scale quarrying. For example, the spoil heaps near Broxburn and Winchburgh in Scotland, locally called bings, are the remains of an oil shale industry that operated between 1860 and 1960. Non-conventional sources, like other forms of mineral extraction, may also be limited by unforeseen local geological factors.

Secondary sources of oil are one reason why some believe that oil production will not fall significantly following the global peak in production. For example, the stated oil reserves of Venezuela, which are mostly made up of heavy oil deposits, are nearly as big as Saudi Arabia. The International Energy Agency state that, "*global oil production need not peak in the next two decades if necessary investments are made*". However, the utilisation of secondary deposits also illustrates why the physical fact of the oil running out, and the price of oil rising, are two very different issues. Oil prices will rise not simply because of a lack of resource. In the first instance it will be because the production of non-conventional oil is more expensive. This is also highlighted by the IEA in their evaluating of world energy resources, stating that, "*resources would not be a limiting factor to supply at the global level... some increase in energy prices, however, may be necessary to stimulate the increase in supply to meet projected demand*" [IEA, 2001].

In the past non-conventional sources have not been viable because they cost more to produce than conventional oil sources. In reality, as the level of conventional production falls, then the rise in oil prices will make non-conventional sources economically viable. However, the underlying assumption in this process is that for supply to be maintained the price of oil will have to consistently increase to fund the development of non-conventional oil sources. Figures from the European Commission [EC, 2001] suggest that with global oil consumption at between 70 to 75 million barrels per day, oil production costs will average $5 to $6 per barrel. But when you move above 75 million barrels per day every extra million barrels of consumption costs an additional $1 per barrel to produce. This large increase in costs is due to sourcing oil from the most difficult to produce or smallest conventional fields, such as Arctic and deep water reserves, as well as production from non-conventional sources.

What is clear from the European Commission's report is that there is a production threshold at which a significant rise in the costs of production must take place. Therefore, even without any speculative economic impacts on oil price, the price will rise purely because more secondary sources are being used. Almost certainly, for a period after the peak in conventional oil production, the total level of oil production will be maintained. This is because the development of non-conventional sources of oil will make up for the decline of conventional oil sources. But even if Venezuela and Canada become more important oil producers because of their large non-conventional oil resources, the large conventional reserves of the Middle East will still dominate global oil production. Even without significant falls in production some organisations, such as the International Monetary Fund (IMF), believe that oil prices will remain at their current high levels until after 2010 [Guardian, 2004d].

The strongest force for driving up prices will of course be a shortage of reserves. Unlike price rises, which are often related to speculation on future supply, major shortages are not likely for perhaps twenty years. It is also possible that the increase in prices, if combined with other political factors that drive prices up further, might reduce the level of consumption. This would reduce demand, perhaps matching consumption more closely to production and making the remaining oil reserves last longer. For this reason some commentator's on Peak Oil talk of a *bumpy peak*, where for a period of time consumption and production interact in a dynamic way as the oil price fluctuates. The effect of this would be to turn the peak into a longer-term plateau.

The key variable in predicting the date of Peak Oil is the global assessment of *total oil reserves* – how much oil is out there to be produced, plus all the production that has taken place in the world to date. In turn, this figure is heavily reliant on how you apply statistical probabilities to the levels of production each oil deposit might produce during its lifetime. Various studies of the date of Peak Oil exist, and these predict a range of dates from 1996 to 2035, depending upon the total oil reserve figure used.

Hubbert himself, in a report published in 1977, predicted a global peak in 1996. This figure was based upon a far higher level of consumption that did not materialise, partly because of the rise in oil prices, and partly because of the fall in demand following the recessions of the early 1980s and 1990s. More recent studies give a variety of figures, depending upon the assumptions made about future consumption and the likelihood of new discoveries. The oil industry researchers Colin Campbell and Jean Laherrère [Campbell, 1998] place the peak in 2005. This is based upon a total global resource of 2,000 billion barrels.

The International Energy Agency (IEA), in their 1998 review of oil resources, put the peak in conventional oil production at 2014, based upon a total resource of 2,300 billion barrels. But in their subsequent 2001 assessment the IEA changed its position. Without any critical evaluation, the IEA followed the assumptions of the US Geological Survey in their assessment of world oil reserves [USGS, 2000]. The USGS cite global resources at 3,345 billion barrels, and put peak production at between 2020 and 2035. Consequently the IEA now put the date of Peak Oil "beyond 2020" [IEA, 2001]. However, the USGS study has been criticised because it places too much emphasis on the discovery of unknown oil reserves, or the development of more efficient extraction technologies.

Getting good estimates of oil reserves is difficult. Almost all the publicly available statistics are taken from surveys conducted each year by the industry's two major publications – *Oil and Gas* and *World Oil*. These two trade journals query oil firms and governments around the world. They then publish whatever production and reserve figures they receive without any validation. As there is no procedure to verify the figures they receive if the companies have problems assessing their reserves, or they assess their reserves wrongly because of the low probability of recovering all their reserves, then the global figure for oil reserves will be inaccurate [Laherrère, 2003].

As accounting standards improve, brought about in part by the collapse of Enron in the USA, companies are being required to prove their reserves more accurately. As with the business practices of Enron, inflating a company's oil reserves can be linked to a desire to increase the market value of a company, its share price, or its ability to produce oil. An example of the problems caused by demands for greater transparency in how oil reserves are forecast is the recent crisis at Shell. They had to re-state their oil reserves in Nigeria 20% lower than their previous estimates [Times, 2004], and in Spring 2004 they were still conducting a programme to try and verify what reserves they have in other parts of the world.

If oil companies must prove their reserves we might believe that this provides more certainty for the consumer, but in reality the opposite is true. This is because most of the world's oil reserves are not owned by corporations, but by sovereign states. So for a large part of the world's oil reserves there is no independent oversight by agencies such as the US Securities and Exchange Commission, and in fact the precise details of these reserves are state secrets.

There is some evidence to query whether or not the overstating of oil reserves by oil producing states has affected the figure for global oil reserves. Six of the eleven OPEC states increased their estimates of oil reserves by between 42% and 197% [Campbell, 1998; Douthwaite, 2003] during the late 1980s. Together these increases totalled 301 billion barrels, representing 15% of Campbell and Laherrère's 2,000 billion reserve estimate, 13% of the IEA's old 2,300 billion barrels, and even 9% of the USGS's very high 3,345 billion barrels figure. For this reason it is argued that any forecasts of world oil reserves cannot be relied upon without some more meaningful process of validation [Heinberg, 2003]. The scale of the change in the reserve figures was startling:

- Kuwait's reserves rose by 26 billion barrels of oil (Bbo) in 1985, an increase of 48%;
- Saudi Arabia's rose by 88 Bbo in 1990, an increase of 52%;
- Iran's reserves rose by 44 Bbo in 1988, an increase of 90%;
- Iraq's reserves rose by 54 Bbo in 1988, an increase of 113%;
- Venezuela's rose by 31 Bbo in 1988, an increase of 124%;
- Dubai's reserves rose by 2.6 Bbo in 1988, an increase of 186%; and
- Abu Dhabi's rose by 61 Bbo in 1988, an increase of 197%.

How much confidence can we have that the government of Saddam Hussein's Iraq were correct in revising their reserves over 113% higher in one year? Possibly little, but if we accept the figures for the world's oil reserves this is what we are asked to accept. Like corporations, states receive benefits from having larger oil reserves. For example, they might find it easier to secure loans from international institutions. The most immediate benefit for these states was securing an increase in their OPEC production quota, which increases the oil revenues received by the state. Be it a state or corporation the temptation is to be optimistic over the level of oil reserves because of the benefits it brings. The oil market does not reward conservative estimates. Therefore, can we trust the figures we are given by the IEA, USGS, and others as accurate?

Another criticism by oil specialists is that both the IEA and the USGS have switched the emphasis in their estimates of oil reserves from *conventional oil* to *non-conventional oil* sources. Within this change in the way oil reserves are assessed exists the potential to overstate the level of world oil reserves because of the limited experience in working non-conventional deposits at very large production volumes.

In particular, the work of the US Geological Survey has been singled out for criticism because it considers reserves which have a lower likelihood of production than those used previously in the assessment of reserves. Instead of a single figure for the proven level of production, as many other studies use, the reserves quoted in USGS studies may have a 95%, 50% or 5% chance of producing the quoted volumes of oil (see Table 2). To put it another way, at a 95% level of probability there's one chance in twenty that this figure *will not be* produced, at 50% it's one chance in two, and at 5% nineteen chances out of twenty (i.e., a one chance in twenty that *it will be* produced). So the law of probability make it more unlikely that the USGS's 3,345 billion barrels figure could be achieved because it considered reserves with a lower

Table 2. The USGS World Oil Reserves Projections
[Source: Adapted from USGS, 2000]

Chance of recovery	Oil and gas liquids, billion barrels			
	95%	50%	5%	Mean
Undiscovered conventional sources	495	796	1,589	939
Improved recovery from conventional sources	205	654	1,102	730
Remaining known reserves				959
World oil production to date				717
Total				3,345

probability of production.

The USGS projections are also heavily reliant on unproven or undiscovered reserves (the first line of Table 2), or the growth of oil reserves through the more efficient recovery of oil (the second line of Table 2). Some of the reserves included are wholly unknown, and extrapolate upon high past levels of new oil finds. Other parts of the study assume that there will be a continued increase in the maximum level of production that can be achieved in conventional oil fields. Together, these factors raise the figure for the total world oil reserve.

Another feature of Hubbert's predictive model is that before the peak in production there should a similar curve and peak that describes the level of discoveries [ECSSR, 2003]. The discovery curve leads the production curve by around thirty years, and, given that the world peak in oil discovery took place during the 1970s, we would expect the production peak to occur soon. One of the key factors in the USGS study, and their 3,345 billion barrels reserves figure, is that an additional 674 billion barrels would be discovered between 1995 to 2025. To achieve this figure 22.5 billion barrels of new oil reserves, on average, would need to be found each year. However, during the first seven years of the study period only 10 billion barrels per year were discovered. To achieve the overall target discoveries over the rest of the period will have to average over 26 billion barrels per year, making the chances that the overall figure will be achieved more unlikely.

The fact that the discovery trend of the USGS study is not borne out by the evidence from recent discoveries, but that the recent data does more closely correlate to models that put the reserves at around 2,300 billion barrels, is another indicator that the USGS estimates are wildly over-optimistic. However, the 3,345 billion barrels figure represents the mean figure from all the sources projected in the USGS study. In other words, there's a 50% probability that this figure will be produced, but there's a 50% chance it will not. Altogether the difference between many other studies of oil reserves, and USGS study, is about 1,000 billion barrels. If you take away the speculative levels of production, and just work at the 95% probability level used by most other studies, the total reserves figure is roughly the same — 2,376 billion barrels. Studies that use this figure for world oil reserves predict the date of Peak Oil around 2015.

The methods of the USGS has led some oil industry specialists to question the motives behind the USGS's methods. For example, in a BBC Radio 4 special report on the oil industry [BBC, 2002a] the researcher Dr. Colin Campbell speculated that political influences on the USGS might have influenced their choice of statistics. As the US is the world's major importer of oil the purpose of quoting such a high figure would be to calm the oil market and keep prices low. This is not an improbable analysis. Recently evidence has emerged to show how political pressure from the Bush administration prevented the US Environmental Protection Agency from taking a stronger line on climate change (this is detailed in the next chapter).

Ultimately all these studies are valid, in that they used statistical models to produce results. Statistical models can be shown to be correct, so long as the sums add up. However, the problem is that if the figures fed into the model are wrong, then the results will be wrong too. The problem with the data used by the USGS is that, unlike the data used by other researchers, it is not borne out by evidence from within the oil industry. In particular, some oil producing states have criticised the forecasts on the basis that they do not have the levels of oil reserves stated in the USGS research [Channel 4, 2004].

In practice, the true level of total oil production is likely not to be known until we are well past the peak of conventional oil production, and the figures demonstrate that the world cannot produce the same volumes of oil as before. Therefore, if we wish to plan for how we as individuals will manage following the date of Peak Oil, we are limited to the data produced by specialists who study oil production. Which set of data you believe, like other forms of value judgement, is largely a matter of which statistical method you believe to provide the most valid approach.

As an example, we could take the recent report by Jean Laherrère [Laherrère, 2001]. He bases his figure for the total world oil reserves, like

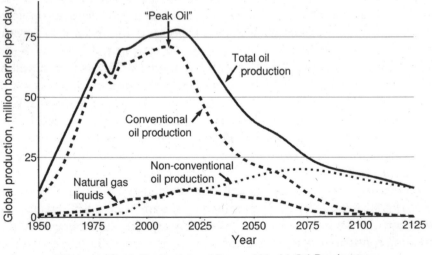

Figure 15. A Projection of Future World Oil Production
[Source: OU, 2003]

Hubbert, on past experience within the oil industry. Figure 15 represents his findings for the total level of oil production, with the separate contributions made by conventional (primary oil), non-conventional (secondary oil) and natural gas liquids (tertiary oil). His data suggests that world oil production will peak at around 80 to 90 million barrels per day (current global oil consumption is in excess of 82 million barrels per day [BBC, 2004b]) around 2010 to 2015. But the precise figure will depend on the oil price. If the price rises in advance of the peak, depressing demand, then a higher production figure might be reached. But if prices see-saw it would produce a *bumpy peak* at a lower level of production. In this case oil might maintain a plateau of production for a few years before any decline in overall production levels emerged.

That may sound straightforward, but the problem for the average person is that different groups of experts, perhaps due to their own motivations or the motivations of those they work for, disagree about how long reserves will last. Some, like Laherrère, use absolute estimates of world oil reserves based on past experience. Others, like the USGS, use statistical methods which mean the figures they give must be interpreted as a probability that the stated quantity will be produced.

To further complicate this equation we have issues such as Iraq. If Iraq can be stabilised, and in the next two years begin to pump five to six million barrels of oil per day, then the peak in oil production will be forestalled for a few years. But if oil production cannot be re-established at this high figure then all Iraq's oil will do is slow the decline in future world oil production, broadening the plateau around the peak.

So, to answer one of the questions posed at the beginning of the chapter, there's a good chance that Peak Oil will occur within ten years, perhaps leading to a plateau in production, and the total volume of oil might begin to reduce in fifteen to twenty years.

Although estimating the date is a significant undertaking in itself, perhaps the greater conundrum will be predicting how the peak will affect the global economy. The greatest impact of Peak Oil on the world economy will be the increase in oil prices. Already there is concern about the rising scale of consumption [BBC, 2004d], and the effect this is likely to have on oil prices in the near future [Guardian, 2004d]. So, how much will oil cost? That's difficult as oil prices are something that energy institutes and governments don't like to speculate upon.

The global oil price is averaged across various forms of oil production – the light crude and the heavy crude – by the way it is traded in the world oil market. In actuality there is more than one oil price, depending upon where it is sourced and traded. The low production costs of the Middle East dominate the market, keeping the price of oil low. But as other sources of oil are exploited then their higher costs must be reflected in the global price of oil. How this happens depends on where this oil comes from, the costs of production, and the costs of transport to the world market. The major oil brands we see in the street – Shell, BP, Esso (Exxon Mobil), etc. – are in fact only minor oil producers at the global level. They make a large part of their income from the import and refining of oil in the industrialised states. In fact there are twenty companies operating around the globe, mostly national or state-owned companies, that produce 60% of the world's oil.

Today there are a variety of factors that influence the oil price:

- *The remaining size of global oil reserves.* In particular, if new information determines that the world's oil reserves are significantly more or less than previously assumed then this will trigger a shift in the oil price.
- *The profitability of recovering oil from different reserves.* The technology of conventional oil extraction is well understood, but there are many uncertainties in extracting oil from non-conventional reserves. It involves mining techniques that are subject to geological conditions, and the greater environmental impact of exploitation might trigger public or regulatory pressures that constrain longer-term development.
- *Changes in the oil price.* This creates an economic feedback mechanism that affects consumption rates. In those states that import a large quantity of their energy as oil, higher prices reduce economic activity, in turn depressing the demand for oil. This is a factor that is highly contentious, and is a factor which the IEA ignore [IEA, 2001] as part of their oil price and production volume assessments.
- *Changes in currency exchange rates.* Oil is usually traded in US Dollars, so exchange rate fluctuations directly affect the price of oil within a state. If the global value of the Dollar falls, because the USA is the largest market for oil, then producers may raise prices to preserve the value of their revenues.
- *Changes to the system of taxation.* Taxes on energy use directly affect the price energy consumers pay, irrespective of oil prices. This is an issue in those states, such as the members of the European Union, that propose to tackle carbon emissions by taxing sources of carbon, reducing their use.
- *Levels of investment in the oil extraction industry.* Oil prices are determined by a range of different production sources. The decision as to whether a particular oil field is economically viable,

and hence the level of investment in producing that source, depends upon the industry's forecasts of future oil prices. This in turn affects the scale of oil production as particular oil fields are priced in or out of the market.

The oil price we see in the media, like the price of most commodities, is made up of various different costs. The price of oil production is only a small part of this. In addition there are transport and marketing costs. The largest part is made up of the income added to the basic price by the oil companies and the oil producing states. It's the level of profit between the basic production price, and the world price of oil, that determines the economic case for the production of a particular reserve. For example, a barrel of crude oil from Saudi Arabia, irrespective of the price paid for it, may only cost $1.60 to lift from the ground and perhaps another $1.40 to bring to market [IEA, 2001]. The rest of the price charged for that barrel of oil goes to the producers and the middle-men. So if the production costs of an average barrel of oil were to rise to $8 per barrel, then the producers and middle men must take a smaller return. Consequently the costs of oil production are critical to the setting of oil prices, and are an important factor in the argument over the increase in oil prices following Peak Oil.

Figure 16 shows the different costs of oil production and supply from conventional and non-conventional sources. Each bar shows the range in production prices for that source. The Middle East has the lowest prices ($2.80 to

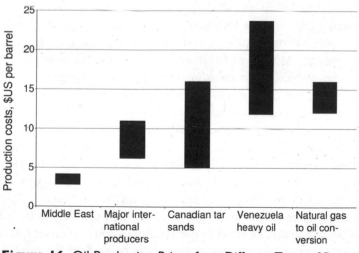

Figure 16. Oil Production Prices from Different Types of Reserve
[Source: Adapted from IEA, 2001]

$4.20 per barrel) because production costs on land are far cheaper, and the large production volumes reduce the overall cost per barrel. The figure for the major producers represents the costs of producing from a variety of environments, both on land and at sea, in hot and cold environments, where production costs are higher ($6 to $11 per barrel). What is more significant are the costs for non-conventional oil. For example $5 to $16 for the most productive tar sands in Canada, or $12 to $23 for the heavy oil deposits of Venezuela. This broader range of prices represents the higher costs of production, and the fact that production volumes are lower than for conventional sources.

As noted in figure 16, it is also possible to produce oil from natural gas, producing a *middle distillate* oil suitable for refining. This requires a more energy-intensive process than the production of liquefied natural gas (the efficiency of gas to oil conversion can be as low as 60% [ECSSR, 2003]) and so the oil produced is very expensive compared to the production of conventional oil. Converting natural gas to oil would also significantly shorten the productive lifetime of the world's natural gas reserves, although for gas producers in areas such as the Far East it is attractive because it allows their gas reserves to be exported more easily.

Oil prices are something that governments take very seriously. Studies by the UK Treasury keep a close watch on the oil price up to a year ahead. These price predictions form one of the key variables in the Treasury's model of the UK economy, upon which future projections of tax and spending are based. It's not just that the oil price raises the levels of government spending. Information produced by the UK Treasury shows that, in the richest seven economies on the globe, the consumer price index (the inflation rate) shadows changes in the oil price [UKT, 2000]. Rising inflation makes everything more expensive, affecting the levels of economic growth [Guardian, 2004e].

As the income of many states is dominated by indirect taxes (like VAT or sales tax) if growth slows, combined with higher inflation, governments don't receive the same value of revenue in taxes. So either spending must fall or direct taxes must rise. Such is the importance of these assessments that the Treasury's decisions on oil price are subject to external review by bodies such as the National Audit Office [NAO, 1999].

The International Energy Agency, as part of their current oil price model, do not believe that oil prices will be affected by any shortage of reserves until after 2020. In their 2001 review, the IEA stated that oil would not move above $21 per barrel until after 2010, after which the higher production costs of the less favourable reserves will raise the price to $28 per barrel by 2020. The IEA's explanation for this increase is the fact that additional investment would be required to extract these more marginal sources of oil. It is also noted by the IEA that during this period the world's dependence on oil would shift to

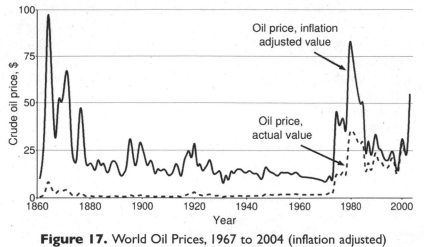

Figure 17. World Oil Prices, 1967 to 2004 (inflation adjusted)
[Source: BP, 2004]

the Middle East. Therefore political influences will have a far greater impact on the world's oil production, and hence oil price.

Obviously, in the light of the $55 and higher prices of early 2005, there must be a problem with the IEA's estimates. Those closely associated with the IEA's position on oil put the blame for this on the operation of the oil market [Guardian, 2004f]. But the fundamental reason for this difference is that those quoting the price of oil between $20 and $30 per barrel are basing their figures on the price of oil production, not its economic value on the world market.

Oil speculation, more than any other factor, has the potential to affect the future price of oil following Peak Oil. As noted earlier, as consumption rises above a threshold of around 75 to 78 million barrels per day the price of production rises disproportionately to the increase in production [EC, 2001]. On current trends, by 2008 oil consumption could be as high as 86 million barrels per day [McKillop, 2003]. This will mean that oil brokers will have to obtain their stocks from more expensive sources, and this would have to be reflected in the global price of oil. If there are problems bringing more oil to the market, then speculation on the future price or availability of oil will drive up the price far faster.

So, how will the price of oil change over the next ten to fifteen years as we move through the date of Peak Oil? That's difficult to answer with certainty, but we could begin by looking at prices in the past. Figure 17 shows the change in oil prices over the last 140 years, adjusted to exclude the effect of inflation. The effect of this is to make the past peaks in the oil price far more expensive when compared to today's oil price. It shows that, except during problematic events in the Middle East, for the last fifteen years prices have fluctuated between $20 and $30 per barrel. For almost a century before this prices were even lower.

The recent stability of the oil price is largely due to the policies of the OPEC states who agree a price range for oil on the global market, and adjust their production levels accordingly to maintain that price. Of course, OPEC has control over the price because they are able to turn the taps on or off of a large proportion of the world's oil supply. The problem is, demonstrated to a lesser extent during the large rise in prices during 2004, as we pass Peak Oil OPEC will be unable to increase production to make the price fall.

Therefore rather than looking at a specific price the question we must ask is, *how far will oil prices rise following Peak Oil?* We need to look at the scale of change in oil prices rather than trying to guess a particular price at a given point in time. Government are reluctant to draw conclusions on future oil prices. Organisations such as the IEA also avoid looking at the external factors to oil prices, such as the effects of economic growth or political instability. This means that the only guide we have are the vague price projections produced by those who observe the oil market:

- The IEA predict a stable price between now and 2010, followed by a 33% increase in prices between 2010 and 2020, purely as a matter of developing production from new sources [IEA, 2001].
- Other IEA-based projections, published as part of OECD reports [OECD, 1999a], estimate a rise of 40% between 2010 and 2015 due to problems with meeting demand.
- If we look at the European Commission's figures [EC, 2001], prices will rise as more marginal sources of oil are exploited. Taking oil production costs as one quarter of the total cost, this would raise prices by at least 25% over the next five years.
- If we take the view of Colin Campbell [Campbell, 2000] as we pass the date of Peak Oil, and organisations like OPEC can no longer keep control over the oil price because of the shortage of oil, prices will jump to around $40 per barrel – a rise of about 75% – purely because of speculation in the oil market.

Given that the causes and effects of Peak Oil are easy to comprehend, we could assume that the rise in oil price following Peak Oil, although steep, will at least be steady. This might not be the case. Increasingly the industrialised nation's tenuous supply of oil may be a target for terrorist action. This is because the economic consequences of even a small interruption to oil supplies could be very large on the world economy. Osama bin Laden, who in particular targets the Saudi state as part of his terrorist action, is reported to have said that the just value of oil from the Arabian Peninsula would be $144 per barrel [Economist, 2001]. Should countries like Saudi Arabia and the other Gulf states experience a change in the composition of their governing elite to one that is less favourable to the West, the effect on the economy would be truly catastrophic.

If we adhere to the standard economic models upon which the globalisation of the world's economy has taken place, the impending peak in oil production creates problems. It's that *First Law of Thermodynamics* again. Cheap oil has been considered by many economists as a passport to economic growth. It enables greater economic activity to take place with little additional cost to the economy. Oil doesn't need a lot of infrastructure to take effect – you can just burn it in boilers and engines and it adds power to the economy. The fact that oil will not be cheap creates problems for economists, and the governments and international development agencies that they work for [NatGeo, 2004].

The wider effect of Peak Oil would be to reduce not just the energy available to industrial society, but also the amount of money that governments have to spend. Economic strife will begin amongst those states who are highly dependent upon petroleum. The higher oil prices will benefit the extraction companies, and states who export a large part of their oil production. However, as the cost is passed down to those at the base of this economic pyramid – the consumer – the reduction in consumer demand will hit jobs because economic activity will reduce as the market contracts. As inflation increases, because goods will cost more, people's income will also become worth less. It's at this point that many governments will experience a major loss of tax revenues because of the reduction in income from indirect taxation (like VAT), combined with demands for an increase in welfare payments and public sector pay rates to meet the rising cost of living.

In the past, as part of any recession, governments borrowed money to finance welfare and re-inflation packages to help the economy recover. Likewise, large corporations could issue bonds, issue new shares, or seek funding for takeovers or consolidation, in order to strengthen their performance during lean times. It's difficult to see how the global financial institutions could lend at this time because there would be a doubt whether those loans could be repaid. Once started the Peak Oil (and a decade or so after that the Peak Gas) recession, unlike other recessions which always had a foreseeable end four or five years ahead, will never end until the economy reorganises itself around more sustainable energy sources. As such a change entails redesigning and redeveloping our modern industrial society, this process might take a few decades (this is examined in detail in chapter 9).

The North Sea is a good example of how Peak Oil will affect the global economy. North Sea oil has had a major impact upon the UK, not just as a source of energy. We could have sourced the oil from other oil producing states and because oil prices take effect globally the price of that oil wouldn't have been significantly higher. What North Sea oil and gas has given the UK is a large amount of revenue from exploration rights and oil production. North Sea oil production, which began in the early 1960s and increased year on year until the end of 1990s, reached the peak of its production – "peak North Sea" – in the year 2000 [Guardian, 2004c]. From now on production levels, even with improvements to the efficiency of extraction, will fall year-on-year by a few percent. It's not that North Sea oil is running out. Half of the extractable oil discovered in the North Sea is still there. It's because, as outlined at the beginning of this chapter, as oil and gas wells approach about half the level of their productive capacity the level of output falls.

What's happening now in the North Sea has been happening for thirty years in the USA [BBC, 2004c] – oil production there peaked in 1970. Whilst self-sufficient in petroleum until the 1960s, today the US imports 60% of its energy needs, mainly petroleum [USDOE, 2004]. Even though North Sea oil has passed its own Peak Oil date, this will not affect oil and gas supplies to the UK. Oil will be imported from South America, West Africa, and especially Russia. Likewise, Britain is the last port of call on a trans-European gas pipeline that brings natural gas from the gas fields of central Asia and northern Russia.

What's more significant is that the revenues from North Sea oil and gas are a significant contributor to the UK's economic health. Putting all the factors together – higher inflation, higher energy prices, and the need to import more energy [BBC, 2004e] – the post-peak oil states, like the UK, face a far wider economic problem as we pass the date of global Peak Oil. Not only will the costs of energy go up, but there will no longer be a large input to the economy from oil and gas production to cushion the impacts of those price rises. In short, there will be less money to go around.

Thus far we've concentrated on oil – after all, this chapter is based on the issues surrounding Peak Oil – but natural gas supplies are also bound by the same geophysical principles. Natural gas will also have its peak, some years after Peak Oil, and according to a recent report from the Parliamentary Office of Science and Technology, this may be as early as 2020 [POST, 2004]. Natural gas will undergo changes to supply, with the accompanying price fluctuations, as oil will do following Peak Oil. When this will be is uncertain. For example, in Europe, the switch to gas from coal for power generation will deplete the gas reserves of the former Soviet Union (our only long-term source of natural gas) far faster.

The North Sea's gas reserves are in decline, and over the next few years Britain will become a net importer of gas. The gas market is far more problematic than the oil market because once local reserves run out, piping gas long distances across the globe makes the supply chain less secure. Turning natural gas into liquefied gas, for transport via ships, is also problematic because 15% is used up by the conversion process, exhausting the gas reserve at a faster rate compared to piped production.

Peak Gas will also affect the way society produces essential commodities. If we look at the information produced by the US Congressional Research Service [CRE, 2001] a doubling in the costs of natural gas (closely associated with the cost of oil) increases the costs of farm fertiliser by about 75% – and fertilisers make up one of the largest costs in modern agriculture, alongside other hydrocarbon-based materials such as pesticides. Of all the various costs associated with food production in the USA, 3.5% are made up of energy costs, 4% are transport costs, and 20% are the cost of the production from the farm. So the scale of long-term changes in energy costs will feed through into the price of food relatively quickly as farmers see the costs of their energy-intensive production methods rise.

As the effects are so serious, you might think Peak Oil would be exercising the mind of politicians. In terms of an open, government-led debate, an in-depth discussion of Peak Oil has yet to materialise. It's not that this issue hasn't been brought into the political arena. For example, in July 1999 there was an all party meeting at the House of Commons on the issue of Peak Oil [Campbell, 1999]. Instead it's more likely that this lack of official interest is due to matters of interpretation. Due to the reports from the USGS, a reduction in oil production is considered a long-term threat – operating at least forty years from the present. For most economists that's the dim distant future. However, as outlined above, the probabilities that the USGS are correct are at best fifty-fifty.

At the European level there has been limited discussion about this matter. The European Commission produced a green paper on energy [EC, 2001] that outlined the problems of oil supply, and noted that Europe only has eight years of indigenous oil supply remaining at current rates of consumption. It also highlights the link between increasing oil prices and the effects upon those individuals and states who live on the margins of poverty. For this reason the green paper stresses the need to *disengage from the oil economy*. Even so, the green paper doesn't consider how soon the world will reach the peak of oil production, and its potential effect on the global economy. The technical annex to the green paper doesn't investigate price fluctuations in any detail either [EC, 2000]. Likewise, the more recent publicity brochure [EC, 2002] on the green paper, produced for the general public to inform them about energy in Europe, failed to consider the economic arguments relating to Peak Oil.

More prosaically, oil production is not considered critical by economists because, it is believed, rising oil prices will price-in other sources to maintain energy supplies. To a certain extent this is true. There will still be a lot of oil around for some time after the date of Peak Oil. However, these alternate supplies will cost more to produce, and it is this factor that will hit the population of the globe the hardest in the first instance – not the shortage of energy that will take place a decade or so after the peak in oil production.

4. Climate Change

When oil is burnt, like gas and coal, it produces carbon dioxide. Carbon dioxide, according to the consensus opinion of many scientists and international institutions, is a threat to mankind greater than the threat of terrorism or war [Observer, 2004b]. Why then, given the problems created by carbon dioxide, is the policy of most governments to burn more oil and gas? If the UK is at the forefront of working to tackle climate change, as claimed by the Prime Minister, why did the UK's Chancellor of the Exchequer demand in the Autumn of 2004 that the OPEC oil states produce even more oil to bring down the price? [Guardian, 2004g] To understand what's going on we need to look at precisely what climate change is, and the problems that might result from our continued use of fossil fuels.

The world's climate is changing. The explanation that most of the climate scientists around the world accept as valid is that the carbon released from burning fossil fuels is creating a thermal blanket or greenhouse that is retaining more energy from the Sun than normal. To begin with we called it *global warming*. Then the scientists pointed out that the extra energy in the atmosphere would create more complex extremes of weather, both hot and cold, wet and dry. So it was re-branded as *climate change*.

Following on from the agreement on climate change made at the Rio Earth Summit In 1992, a large numbers of scientisrs across the globe have been awarded funding to study climate change and produce better computer models of the atmosphere to predict its effects. The fact was that whilst the theory stood up, there just wasn't enough data about the Earth and its natural systems to decide how adding more greenhouse gases would change our climate. So much of the last decade has been spent measuring the Earth's climate, and the impacts of our industrial and agricultural activities on natural systems, to get enough data to decide how climate change will affect different parts of the globe.

As well as looking at past climatic changes from fossil records, scientists have also been looking at the changes in climate today. Changes in the rates of plant growth, the affects of warmer temperatures on marine species, and the changes in temperature and rainfall. Scientists have also been studying the oceans, and in particular the effects of climate change on ocean currents. It was whilst studying the *North Atlantic Thermohaline Circulation*, warm water current that flows from the Caribbean north across the Atlantic to the coast of northern Norway, that oceanographers discovered a new twist to the climate change theory for residents of north west Europe – *abrupt climate change*.

Abrupt climate change raises the possibility that over the next twenty to fifty years, over a transition that might take just one or two years, the Gulf Stream and the heat it provides might just *switch off*. This means that unlike ordinary climate change, where we increasingly sweat over the course of a century, north west Europe would experience a rapid cooling to a level where it experiences a climate more representative of its latitude – that of northern Canada and Alaska. So, whilst many people were thinking that if Surrey had the climate of the south of France then 'climate change was OK by them', in fact, we're heading for a climate more like that of Newfoundland.

Abrupt climate change is not a new phenomenon, but it's only recently we've understood the mechanisms by which it can take place. It has happened a number of times over the past 400,000 years, and the evidence of these changes poses some difficult questions for European governments. This is because just at the time when northern Europe will have problems finding energy, as first oil and then natural gas production goes into decline, there is a very real possibility that they're going to need a lot more to make the colder climate more tolerable.

A significant problem about tackling climate change is that it entails a major rethink of the lifestyles of those inhabiting the industrialised nations. For this reason many politicians are also sceptical that the solution offered by scientists – significantly reducing the carbon emissions from energy use – is achievable. It's also important to realise, when studying the messages we see in the media on climate change, that some of the solutions offered to climate change are not as sound as they could be. For example, the issue of climate change makes the nuclear industry beam with delight because it allows them to promote their distrusted energy solution as a low carbon technology. However, on balance, it's arguable that nuclear won't produce a secure source of energy for the foreseeable future.

The political uncertainties over climate change mean that, in countries such as the USA and other states that are rich in petrochemicals, industrialists have donated money to sceptical scientists in order to challenge the theory of the greenhouse effect's role in climate change. As is explained later in chapter 10, the extra profits that these countries or large corporations can gain by forestalling action over climate change makes giving money to those opposing the scientific basis for climate change a highly cost-effective investment.

For some people the idea of climate change is nonsensical. How can the tiny inhabitants of this huge planet affect the atmosphere in this way? The amount of energy we use is nothing compared to that the Earth receives from the Sun, so how can we cause this effect?

Compared to the average person the Earth is huge, but it's not the size of the planet that matter's, it's the size of the atmosphere. The diameter of the

Earth is around 12,750km, but the thickness of the entire atmosphere is just 100km. When you look at a tomato, if the Earth is the body of the tomato, then the thickness of the atmosphere around the Earth is about the same thickness as the tomato skin. And, in volume terms, the atmosphere, where the weather is, is only a few percent of the total volume of the whole planet.

The fact that we can change the climate is not just because the atmosphere has a comparatively low volume. Natural systems keep the climate of the Earth within a narrow band of temperature, moisture, and with a certain composition of atmospheric gases. This is a process known as *homeostasis*. In the same way that a central heating controller knows when to turn the radiators on or off to keep an even temperature, so the climate systems of the Earth balance the natural systems through a highly complex set of physical and biological cycles. However the Earth hasn't just got an on switch like a thermostat. It has a number of off switches too. To regulate climate the atmosphere relies upon a series of related natural cycles that provide *positive feedback* (an amplifying effect) and *negative feedback* (a dampening effect). Together they maintain the state of the atmosphere and the climate. Some liken the way the Earth's atmosphere regulates itself to the way an organism regulates its metabolism [Lovelock, 1979]. The problem is that, as we are on the inside of this organism looking out, we can't fully understand the way that whole organism works. This makes it difficult to understand how our impacts on climate will work out over time.

As noted in chapter two, the Earth has certain energy inputs. The natural cycles that maintain the climate use these energy inputs to operate the various energy and chemical balances that maintain climate. The Earth's atmosphere is composed of a series of layers, and the gases in those layers affect the way that the energy from the Sun is absorbed. 30% of the Sun's energy is reflected back into space as long wave infra-red radiation. So at the crudest level, by managing how much energy is reflected into space, the Earth can regulate how much of the Sun's energy it receives. The level of energy reflected back to space is influenced by two factors:

- The albedo of the planet – if the planet is darker it absorbs more energy, if it is lighter it reflects more, so increasing the amount of cloud or snow cover increases the reflection of heat, and vice-versa.
- The atmosphere and the Earth's surface trap the Sun's energy as chemical energy or heat, and one of the key regulators in this process are the greenhouse gases – such as carbon dioxide, nitrogen oxides and methane. These makes the planet warmer than it would otherwise be, and consequently it can support a

Box 9. Climate Change and the Earth's Orbit

The Earth orbits the Sun in an ellipse. Although the average distance is 93 million miles, when the Earth reaches perihelion (the closest point to the Sun, roughly January 1st or 2nd each year) it is slightly closer than the average. When it reaches aphelion at the end of June, it is slightly further away. This varies the energy input, giving more energy to the southern hemisphere.

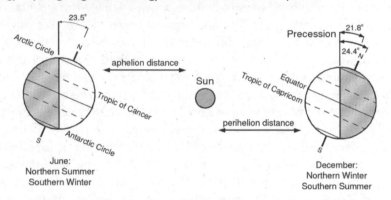

Figure 18. The Earth's Orbit and Variations in Solar Radiation

The axis of the Earth, around which it spins, is tilted away from the plane of the solar system. This is what creates the change in seasons in the northern and southern hemisphere. It also means that for a large part of the year either the north or south polar region is kept in complete darkness or daylight. Currently the angle of tilt is 23.5° from the plane of the solar system. This means that at perihelion the Sun is overhead at 23.5° south of the equator – the *Tropic of Capricorn*. At aphelion the Sun is overhead at 23.5° north of the equator – the *Tropic of Cancer*.

The Earth's elliptical orbit has a small eccentricity. This slowly changes the perihelion/aphelion distance of the orbit over a cycle lasting 400,000 years. The tilt of the Earth on its axis also wobbles. This precession of the tilt of the axis works over a cycle lasting 25,800 years, during which the tilt varies between 21.8° and 24.4°. This results in the position of the equator, the tropics and the poles moving slightly, but in combination with the Earth's orbit the precession creates a cycle of changes to the tropics that operates over 100,000 years. These cycles work together, varying the total energy the Earth receives depending on the joint effect of the 400,000 and 100,000 year cycles. This variation in the energy received by the Earth is called the *Milankovic-Croll effect*. It was proposed by a Scottish astronomer, James Croll, in the 1860s, and confirmed by the astronomical observations of a Yugoslav astronomer, Milutin Milankovic, during the 1920s and 1930s.

more diverse range of life.

The Earth's albedo and greenhouse gas levels represent a large scale equilibrium that controls the temperature of the Earth. How the levels of greenhouse gas, and the levels of cloud or snow cover that affect the albedo, operate is far more complex. These complex cycles are inextricably linked with the existence of biological organisms on the planet. Although the planet's natural systems seek to operate within a relatively narrow range of climatic conditions, over the history of the planet the climate has varied enormously. However, as these large changes in climate take place relatively slowly, over cycles of 100,000 to 400,000 years, life can learn to live with the altered conditions.

The underlying flaw with a climate that is inter-related with life on Earth is that any process not related to the biological systems within the Earth's atmosphere can disrupt the natural balance of this system. The main agents for change, excluding man, are volcanic eruptions, changes in the configuration of the tectonic plates that make up the surface of the Earth, and asteroid strikes. By changing the equilibrium point of the atmosphere these factors can make the Earth's climate swing between cold and dry, to hot and wet, over a short period of geological time – perhaps only a few decades. For example in 1815 a volcano in Indonesia, called Tambora, underwent a massive eruption. Across the world in 1816 temperatures fell as the volcanic ash in the atmosphere lowered the input of solar radiation. This is a complex set of reactions, but put simply, the dust and gas aerosols released by volcanoes cause more clouds to form, increasing the Earth's albedo and so lowering temperatures at ground level. In Europe, 1816 was known as the "year with no Summer", immortalised in Lord Byron's poem, *Darkness* [Byron, 1816].

A major influence on the Earth's climate are cyclical changes to the Earth's orbit. When you look at a map of the world you see the poles, the equator, the tropics, and the arctic circles (see Box 9). However these lines that we drawn with such certainty on maps are, in terms of geological history, not fixed. They move around, and this variability affects the climate. One consequence of this variation are the ice ages. An ice sheet grows over a period of 90,000 years whilst the average temperature slowly falls, and then melts over the next 10,000 years when it rises again (we are currently at the beginning of one of these 90,000 year cooling periods following the last ice age that ended around 10,000 years ago).

Another impact on climate is the movement of the plates that make up the Earth's crust. Over millions of years the continents have moved around, broken up and fused back together. In the short-term this is significant because it causes volcanic activity. In the longer-term the full effect isn't wholly understood, but it appears to create significant cooling on the Earth when at least one of the continents is located over a polar region (currently, Antarctica). If not, the Earth is a lot hotter, wetter, and the sea levels are higher. However, in terms of human history, the movement of the Earth's crust isn't really an issue because of the large time-scales over which it operates — far longer than the variations in the Earth's orbit.

The climate of the Earth millions of years ago can be inferred from the type of fossils we find, or the types of sedimentary rocks that were laid down during those periods. To get a realistic assessment of how the climate is changing we need to have a reliable measurement of the actual temperature and gas concentrations, for different parts of the globe, for the last tens or hundreds of thousands of years. Weather records exist for some European states, like the UK, as far back as the Seventeenth Century, but the most reliable direct measurements of the Earth's temperature and atmosphere only exist for the last century. To go back further than that we have to use natural indicators of climate called *proxy data* (see Box 10). By combining proxy data from many sources it is possible to produce a picture of what the global climate was like in the past. This allows us to examine changes in the Earth's climate over the past half a million years. Other proxy data provide information over millions of years, but they are not as accurate.

One of the ways the climate has been monitored across the globe is by taking samples of deep water sediments. The rivers of the world collect organic matter and wash it into the sea. This material is then fed upon by plankton, which then die and fall to the bottom. As more sediment falls out of the water, layer upon layer of sediment is compressed and preserved. By drilling into the sediment from a ship a sample column of the preserved sediment can be removed and examined. Some tests look at trapped minerals. For example particles of volcanic ash give clues to regional volcanic activity. The best guide to climate generally is by studying the remains of the plankton, both the diversity of species present and what their shells are made of — a field called *micropalaeontology*. From studying the plankton and microfossils of today and the environments they live within, and by identifying the ancestors of modern plankton, scientists are able to identify the climate at the time a particular layer of sediment was deposited.

The other major source of proxy data are ice cores. Using a similar method to producing cores of sediment, ice cores are taken from the world's major ice sheets in Antarctica and Greenland. These are also analysed, but instead of looking for microfossils, scientists look for small bubbles of gas frozen in the ice. As snow falls it traps air, and as the snow compresses to form ice this ancient sample of the Earth's atmosphere is trapped. By analysing the gases in the ice core we can get readings of gas concentrations in the atmosphere, and estimates of the temperature at that time. For example, one of the indicator

Box 10. How Climate Proxy Data is Collected

The age of sediments can be tested using radiocarbon dating. So by identifying a layer with microfossils indicative of a certain climate, dating those fossils will tell you what the climate was like at a certain point in time.

As carbon dioxide circulates high in the atmosphere cosmic rays change the atomic structure of the carbon atom from ordinary carbon-12 (^{12}C) into radioactive carbon-14 (^{14}C). The ^{14}C rains down from the upper atmosphere and is taken up by all living things. All living things have a roughly constant value of ^{14}C in their bodies, but after death the old ^{14}C is no longer replaced with new ^{14}C, so the level of ^{14}C begins to decay – halving in quantity every 5,700 years. By measuring the concentrations of ^{14}C in old organic material, and because we know how much there was to start with, we know how long it has been dead.

Another indicator taken from the plankton in sediments is the ratio of two isotopes of oxygen: oxygen-16 (^{16}O), which makes up about 99% of all oxygen, and oxygen-18 (^{18}O), which makes up most of the rest. ^{18}O is heavier than ^{16}O. This means that ^{18}O evaporates less quickly from sea water than ^{16}O, and it condenses more quickly to form rain and snow than ^{16}O. So there should always be more ^{18}O in the seas compared to the rest of the environment. What this means is that the amounts of ^{18}O in seawater, and then trapped in ancient sediments, varies according to how much of the ^{16}O has fallen in the sea to dilute it. The larger the amounts of snow and ice present on land, the less ^{16}O is available to dilute the ^{18}O in the seas. So if the ratio of ^{18}O to ^{16}O in sediments is high, it means that more ^{16}O is locked up in snow and ice, and consequently that the climate of the Earth must be colder. By combining the ^{18}O to ^{16}O ratio for a particular layer of sediment with a value for the ^{14}C concentration, the temperature reading can be dated.

A similar procedure is used to date ice cores. The ratio of ^{18}O to ^{16}O is of less value since most of the ^{18}O is found in the sea. Instead the same type of measurement is taken using an isotope of hydrogen called *deuterium* (^2H). This, like ^{18}O, produces a measure of temperature. In order to date the information from ice cores the data for temperature is mapped to the data produced from sediments.

By measuring the thickness of tree rings it is possible to estimate the length of the growing season for that year – hence how warm the weather was. The results of tree ring analysis can also be mapped to sediment and ice core data to produce a more comprehensive picture of global climate variation – except that tree ring data gives a reading for the climate in the area where the tree grew rather than other forms of proxy data which work at the regional or global level. Another useful aspect of tree ring data is that the levels of ^{14}C raining down from the sky are not always constant. When the Earth's magnetic field weakens, more cosmic rays get through, creating higher levels of ^{14}C. By using tree ring data these anomalies in the ^{14}C record can be eliminated, improving the accuracy of radiocarbon dating.

gases that climate scientists look for is methane. As well as being a greenhouse gas associated with heating of the atmosphere, it is a measure of how much marsh and bog was present on the Earth, and so indicates how hot or wet it was in the past. The ice also traps microscopic particles of dust from the atmosphere which allows scientists to look for evidence of volcanic eruptions, large forest fires, or the dust from dust storms created when the world is hotter and dryer.

Another widely used source of climate data are tree rings – the science of *dendroclimatology*. The oldest trees on Earth are perhaps 2000 years old, but by adding to this samples of wood preserved in bogs and marshland tree ring data can provide a measure of the climate over tens of thousands of years.

The proxy data used to work out past temperature changes consists of tens of thousands of measurements taken from ice cores, sediments and tree rings. By putting the all the proxy data together we get a comparison of global temperature versus carbon dioxide concentrations for the last 425,000 years (see figure 19). The data shows a clear pattern. The temperature (the dotted line) and carbon dioxide (the solid line) measurements correlate. If one goes up, the other goes up in nearly the same proportion, and vice-versa. Also, within the 100,000 year cycles, the carbon dioxide concentration and temperature start off high, and gradually fall away. This is the cycle created by the precession of the Earth's axis, but within this regular 100,000 year cycle are smaller variations. In fact, figure 19 has been simplified, removing the noise from the data to better show the correlation of temperature and carbon dioxide. The actual data set contains many small peaks and troughs over short spaces

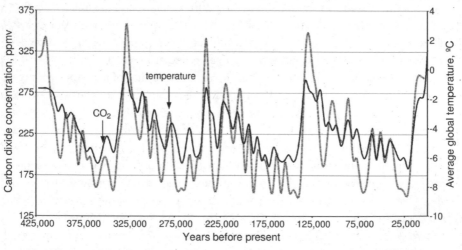

Figure 19. Climate and Carbon Measurements from Proxy Data
[Source data: RCEP, 2000]

of only a few thousand years. There are various explanations for these rapid changes, abrupt climate change being one of them.

The fact that temperature and carbon dioxide levels have varied synchronously for the last 400,000 years is a good indication that changing the carbon concentrations of the atmosphere today will have an impact upon global temperatures. Over the last 400,000 years, up to about 200 years ago, carbon dioxide levels have not risen significantly above 300 parts per million by volume (ppmv). In fact, for most of the last 2000 years carbon dioxide levels have been hovering around the level that would suggest that the cyclical trend of the last 400,000 years was going to repeat again. This changed around 1800 when the large-scale use of coal, begun in the early stages of the industrial revolution one hundred years before, was beginning to have an effect on the atmospheric concentrations of carbon dioxide. By 1900, carbon dioxide levels were exceeding the maximum levels of the last 400,000 years. Today levels have reached 370ppmv, a level of carbon dioxide not seen since the Pliocene period 3 million years ago.

What climate scientists are concerned about is that if we burn most of the oil and gas, then carbon dioxide levels could rise beyond 500ppmv – a level not seen on Earth since the Eocene period 35 to 57 million years ago. Such a level is forecast to increase global temperatures by 4°C. If we burn the remaining coal in order to make up for the loss of oil and gas, then carbon dioxide levels will rise up to or past 1,000ppmv – a concentration not seen since the Cretaceous period 70 million years ago (when dinosaurs were the dominant species on the planet). Carbon dioxide levels this high could raise global temperatures by at least 6°C. This could lead to other climate hazards, such as widespread forest fires or the release of trapped methane from the oceans, that would push carbon levels higher still. This runaway in climate, as global temperatures rise 10°C or 12°C, might create the kind of global extinction scenario that happened at the end of the Permian era, 250 million years ago, when a large part of the world's land masses turned into deserts.

It's important to note, even though we are still only at 370ppmv, these predicted temperature rises are not happening at some time in the future – they are happening now. According to the International Panel on Climate Change, 1998 was the hottest year for which direct temperature records exist, 0.55°C warmer than the average temperatures recorded between 1960 and 1990. In global terms, the ten warmest years on record have all occurred since 1990, and the Twentieth Century experienced the greatest level of warming than any century over the past 1,000 years [BBC, 2004f].

Today most international organisations accept that climate change exists, and that it is due to *anthropogenic* (human created) emissions of greenhouse gases. Even so, the complexity of the Earth's climate systems is still not fully understood, and this may mean the results of current predictions may not be wholly accurate, or may underestimate the effects of climate change. For example, scientists cannot explain the mechanism that produced the recent sharp rises in global carbon dioxide levels [Guardian, 2004h].

Despite the large body of evidence that exists, US-based researchers linked to the pro-business lobby group The Marshall Institute [Marshall, 2004] have questioned the whole basis of the use of proxy data. The researchers, who are astrophysicists, argue that the types of warming predicted by climate change models cannot be substantiated from current data [Climate, 2003]. However, they have stated that given the available instrumental data, rather than proxy data, then climate has warmed significantly during the Twentieth Century as carbon dioxide levels have risen [BBC, 2004f]. More significantly, data from recent satellite surveys, which has provided a qualitative difference in the data we have on the level of warming around the globe, tend to confirm the IPPC's projections [Guardian, 2004i].

Although climate change hit the news towards the end of the 1980s, it was proposed as a possibility by two researchers at the Scripps Institute of Oceanography in the USA, Roger Revelle and Hans Suess, in 1957. As the body of evidence for climate change grew during the 1980s, in 1990 the UN's *Second World Climate Conference* decided that an international agreement limiting greenhouse gases was required. The *Rio Earth Summit* in 1992 was the launch pad for many global initiatives on the environment. Climate change was one of these. Also, as part of the conference agenda, participants signed up to the *Rio Declaration* [UNCED, 1992]. *Principle 15* of the Declaration states that,

> "In order to protect the environment, the precautionary approach shall be widely applies by States according to their capabilities. Where there are threats of serious or irreversible damage, lack of full scientific certainty shall not be used as a reason for postponing cost-effective measures to prevent environmental degradation."

It was on this basis that one of the treaties agreed at the Rio conference was the *Framework Convention on Climate Change*. Under the auspices of the United Nations, this agreement set up a process whereby states would develop a framework for reducing greenhouse gas emissions. As part of the effort to co-ordinate a global research programme on climate change the UN had previously convened the Intergovernmental Panel on Climate Change. This is now the leading body on climate change, providing scientific assessments from more than a thousand researchers across the globe.

The core of the Framework Convention on Climate Change is a requirement for the developed nations to reduce their greenhouse gas emissions.

Whilst the Framework Convention sets the objectives, the detail of how these objectives would be achieved was thrashed out in the *Kyoto Protocol*, agreed in 1997. With the accession of Russia in late 2004, enough nations have now signed up to the Kyoto Protocol so it can now become part of international law, and from February 2005 those nations who have signed it are bound to implement its provisions. Under the Kyoto Protocol the UK has agreed to reduce carbon emissions to 12.5% below the 1990 discharge level, and this target should be met between 2008 and 2012 [DTI, 2004d].

The problem is, despite the delay, the level of cuts mandated by the Framework Convention on Climate Change will not be enough to avert serious climate change. As recently admitted by the IEA, even with controls on emissions carbon emissions are likely to be 40% higher at the end of this decade [Guardian, 2004j]. The ineffective nature of the Kyoto Protocol is one of the challenges made by those who oppose the treaty, in particular the Danish statistician and game theory scientist, Bjørn Lomborg. In his book, *The Skeptical Environmentalist* [Lomborg, 2001], he observes that The Kyoto Protocol, as agreed, will only reduce the heating effects of climate change by 6%. Lomborg argues that there are equally pressing problems facing the globe apart from climate change [Guardian, 2004k], an idea he recently brought together with others in what is called *The Copenhagen Consensus* [Lomborg, 2004]. However, the validity of the environmental case he advances to support this position has been challenged and found lacking [Guardian, 2004l] because the types of cost–benefit analysis Lomborg uses are based on previous valuations of economic indicators, not the long-term scientific case for why we have to change industrial society (the issue of economic justifications for pollution is examined later in chapter ten).

The problem is that Kyoto has been the only process around to tackle climate change and therefore many environment groups have supported it even though its effects in slowing climate change will be minimal. Environmentalists have supported it precisely because there is no other realistic mechanism for securing global reductions in carbon emissions. However, this does not mean that Kyoto is the correct approach. As identified by the UK government's panel on environmental issues, the Royal Commission on Environmental Pollution, to have any real effect greenhouse gas emissions need to be cut by much, much more. Their recommendation was that the UK cut carbon dioxide emissions by 60% by the year 2050 [RCEP, 2000]. This is the point made by those who promote more drastic solutions, such as the *Contraction and Convergence* programme [Meyer, 2003] promoted by the Global Commons Institute [New Scientist, 2003].

For once, Britain was quick off the mark on climate change. It produced its proposals for meeting the targets set in the Framework Convention in under two years [HMSO, 1994], and in fact, Britain was one of a few countries that actually met the Framework Convention's target for the reduction in greenhouse gas emissions. However, the reason Britain met its commitments under the Convention was almost entirely due to the switch from coal-fired electricity generating plants to new gas-fired plants during the late 1980s and early 1990s. As will be discussed in the next chapter, this reduction in the UK's greenhouse gas emissions is only temporary, and we'll be back above the 1990 emission ceiling in around a decade [DTI, 2002c].

Amongst the public at least, the climate change debate seems to be centred on the effects of a hotter climate and the inability of politicians to deal with them. It's ironic that the UK's position on climate change was originated with the strong support of Margaret Thatcher, a traditional Conservative, who in a speech to the Royal Society outlined the threat in terms of experimenting with the environment. Following on from this traditionalist approach, the public debate on climate change in the UK seems to be linked to the effects on wildlife rather than on the human population. For example, the recent studies, reported widely in the UK media, claiming that as climate change accelerates we are already at the beginning of a new mass extinction during which more than one million species will become extinct [Guardian, 2004m]. At the European level studies of the impacts of climate change tend to take a slightly wider view on the impacts, both on human society and on wildlife [EEA, 2004].

If we take a critical analysis it would appear that the UK's mainstream environmental groups are unwilling, or unable, to comprehend the debate in terms of human populations. Rather like the discussion promoted by Bjørn Lomborg, as to whether the Kyoto process is adequate because its impact is so minimal, examining climate change in terms of the UK's human population is perhaps too politically challenging as it involves talking about the unpleasant consequences of human activity (see the next chapter). However, the reality is that, rather like the discussion of Peak Oil, the only way to achieve a meaningful resolution to the threat of climate change is a root and branch revision of the way our society operates.

In contrast, the public debate in the USA does not focus on climate change as a force detrimental to nature. The debate on climate change is almost wholly centred on the effects controlling greenhouse gas emissions will have on the American economy. This is in part due to the American controls on political donations that ban corporations from directly funding politicians, and which skew American political debate towards the needs of large interest groups. However, under President George W. Bush, this process has changed. No longer do lobby groups, such as the American Petroleum Institute, need to argue from the outside. When George W. Bush needed staff to run his administration he sourced many of them from within the community of pro-business think tanks. One of these pro-business groups, The Heritage Foundation, gave George Bush a list of people he should appoint to run his administration. During his first administration he employed over 200 people from the list [BBC, 2002b].

It should be no surprise then that there have been various allegations that Bush administration appointees to the US agencies responsible for environmental matters have been distorting the presentation of issues like climate change, forestalling the need for action. Criticisms of pro-business/anti-Kyoto Protocol bias within the Bush administration began with the appointment of managers to the White House Council on Environmental Quality (CEQ). They had links to law firms representing large industrial corporations, and to the American Petroleum Institute. What brought this matter to the public's attention was when a staff member at the US Environmental Protection Agency (USEPA) leaked a memo alleging that the White House, via the CEQ, had intervened to prevent detailed information on effects of climate change being included in the USEPA's annual report [BBC, 2004g]. The memo stated:

> "The White House has made major edits to the climate change section of the EPAs report on the environment. Indicating that, quote, no further changes may be made... The section does not address the effect of climate change on human health, and the environment. This will be conspicuously different from other chapters of the report... Most important, the [Report on the Environment] no longer accurately represents scientific consensus on climate change." [USEPA, 2003]

This leak, and the implications for the climate change policies of the White House, were covered widely in the US media [WP, 2003], and also the UK's broadcast [BBC, 2004g] and print [Observer, 2003] media. Relations between US politicians and the international bodies studying climate change descended still further when the Republican chairman of the Senate Environment Committee described climate change as "the greatest hoax that has ever been perpetrated upon the American people" [BBC, 2003a].

Despite this block within American politics to any movement on climate change it seems that there is some movement from within the US defence establishment. A Pentagon report [GBN, 2003], commissioned by the US Defense Department's senior adviser on global security Andrew Marshall, and authored by a former CIA advisor and a leading US businessman, warns of the potential dangers that climate change poses to global security [Observer, 2004b]. Even so, despite scandalous leaks and pro-industry posturing, it's clear that no development on greenhouse gas controls will happen at the international level until there is a serious realignment within the American political establishment.

One of the problems with the debate on climate change is that, like the debate on Peak Oil, it's just too far off the horizon of most politicians to really impact their political judgement. That may change if evidence from recent oceanographic studies is accurate. As noted earlier, in relation to past global temperatures, the climate has a tendency to abruptly see-saw over long periods and over very short periods, as little as a few decades. In terms of an atmospheric effect this is difficult to explain. Some speculated that it could be the effect of longer term volcanic activity, but it was clear from the fossil record that the climate would swing, globally, for a period of a few decades, or centuries, within the space of a few years. What has recently been discovered is that the world's oceans, rather than being a passive participant in climate change, play a major role in how heat is moved around the planet. What controls this process is the salinity of sea water. If you change that salinity then the whole system breaks down, and the result is the complete opposite of what many in the developed world think climate change is all about.

The warm ocean surface currents, heated by the Sun at the equator, become buoyant and move towards cooler water. As so much water evaporates around the equatorial region the water there is more salty. Salt water is heavier than than fresh water, but because the tropical salt water is warmer it is more buoyant and floats upon the surface of the fresher but colder temperate zone sea water. When the warm water reaches higher latitudes it dumps its heat into the atmosphere and cools, losing its buoyancy. It then falls under its own weight to the bottom of the ocean and circulates back southward.

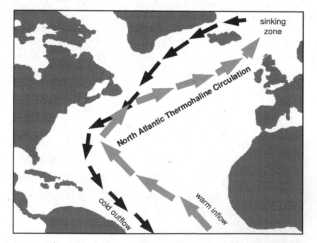

Figure 20. The North Atlantic Thermohaline Circulation

The warm currents driven by the salinity of seawater are called the *thermo-haline circulation*. The network of thermohaline circulation currents around the globe are know as the *ocean conveyor*. The conveyor extends from the Pacific Ocean, above Australia into the Indian Ocean, and then around the tip of Africa where it flows north into the Atlantic. The conveyor is crucial to climate because it helps transport about 50% more heat from the equator to higher latitudes than would otherwise be possible through ordinary ocean currents [WHOI, 2003].

Our local leg of the conveyor, the *North Atlantic Thermohaline Circulation* (or NATC – not to be confused with the Gulf Stream which is a maritime current) carries the heat energy of a million power stations. It means the citizens of Europe can swim in the sea at the same latitudes that Canada has ice and polar bears. But the NATC forms one end of the ocean conveyor, and therefore any change in the NATC's strength has global implications. Evidence from oceanographic studies of how the thermohaline circulation operates, and from ice cores and sediments from the North Atlantic area, seem to indicate that the sinking of salt water in the area between Greenland, Iceland and Norway is a finely balanced process. A one percent change in the salinity may be significant enough to make the water at higher latitudes too fresh to sink. Without sinking, the force that draws water across the North Atlantic from the east coast of the USA would disappear – the conveyor would turn around in the mid-Atlantic and go back again.

The problem today is that climate change is causing more fresh water to enter the North Atlantic. The Greenland ice sheet is melting. Fifty cubic kilometres of ice and snow are running-off Greenland as freshwater each year. This is double the rate it was ten years ago [BBC, 2003b]. Rising temperatures are also causing increasing run-off from the three major rivers that drain northern Russia and Siberia allowing an additional 128 cubic kilometres per year of freshwater to flow into the North Atlantic. Increased ice melt in northern Canada is also bringing more fresh water into the North Atlantic via the Labrador Sea. Measurements of salinity in the deep currents around the North Atlantic have shown a steady decrease in salt content for the last forty years, but the process has been especially pronounced over the last ten years [Dickson, 2002].

Research from proxy data has shown that in the locality of the North Atlantic there have been a number of periods over the last few thousand years where temperatures have dropped significantly – a process now called *abrupt climate change*. Changes in temperature do not happen gradually over decades, but over the space of a few years. About 12,700 years ago, in a period called the Younger Dryas, the temperature dropped 5°C and remained low for the next 1,300 years. 8,200 years ago another shorter cooling, of about the same magnitude, lasted for around 50 years. The most recent event was

deliberations over energy policy, and decisions over large-scale energy developments, are taken centrally also means that political pressure from the public is difficult to apply. It's not just that the decision-makers are hard to get at. This closed-door environment ensures that industry lobby groups can work privately behind the scenes – a fact illustrated recently when Tony Blair admitted that US-based nuclear lobbyists had convinced him to keep the nuclear option open in the UK [Guardian, 2004n] against the prevailing public opinion.

The UK's gas, electricity and nuclear energy companies, privatised during the 1980s and early 1990s, are regulated by the DTI, as is the development and regulation of renewable energy sources. As part of this work the DTI compiles data on energy use in the UK, and awards funding for research projects on energy use or the development of new energy sources. The DTI also conducts studies of future energy use, and it publishes these as a series of energy reports and statistical bulletins. The most recent comprehensive study of the UK's future energy demand is *Energy Paper 68* [DTI 2002c]. It sets out the likely changes to energy demand over the next twenty years, how this energy will be produced from different energy sources, and the change this might cause to the level of greenhouse gas emissions. These figures are likely to be reviewed in the light of the Government's decision to revise the UK's climate change programme [DEFRA, 2004].

The EP68 energy model often produces a slightly lower figure for energy use than is shown in the real collected information. This is due to the differences between the EP68 model and the way the real energy statistics are collected. For example, the total energy demand is projected to rise from 218.7mtoe in 1995 to 246.5mtoe in 2010, but in 2002 the real figure was 237.7mtoe [DTI, 2002b]. When reading EP68 it's important to look not at the actual figures, but the proportion of the rise or fall that the model predicts.

Energy Paper 68 is complex to interpret because it runs different energy models in parallel. The models are based on different levels of economic growth – *low* (2%), *central* (2.5%) and *high* (3%) – and different levels of energy prices – *low* and *high* (the prices varying according to fuel type). This creates six different scenarios for energy use. The models give estimates of energy use overall, and within different energy sectors – domestic, industrial, transport and services. Figure 21 shows the projected growth in demand on the 'central' growth forecast, including the effect of both high and low fuel prices – which produces a band of possible values rather than a single line.

The central growth scenario assumes that economic activity rises 2.5% per year. This causes energy use, on average, to increase by nearly 13% over 15 years. If growth is higher or lower than this, then demand will be higher or lower than the band indicated on the graph. Industrial demand has the least

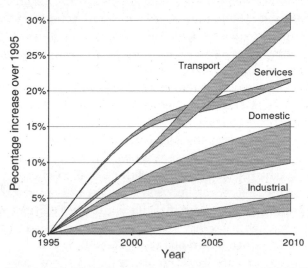

Figure 21. EP68 Energy Demand Forecasts, 1995-2010
[Source: Adapted from DTI, 2002c]

growth – between 3% and 6%. Domestic growth is between 10% and 15%, and service growth is around 22%. The largest increase in demand, between 29% and 32%, is in the transport sector. Most of the growth in the transport sector is for road fuels – petrol accounts for 39% of the total, and diesel 35%. Aviation fuel constitutes 25% of the total figure.

The interesting fact about DTI forecasts is they tend to be an honest appraisal of what the business and energy sector want to do in the future. After all, it is the purpose of the DTI to help facilitate the growth of UK industry, and so a large part of what they do is guided by the interests of business. But these growth figures are in sharp contrast to the statements from the Department of the Environment. In this Department the emphasis is that the growth in transport, and energy use in general, needs to be curbed. This creates a clear difference in planning and policy between the two departments when it comes to the issue of managing the impacts of climate change. In fact, the more you look into the issue of energy, from policy on nuclear power to the development of renewable energy, this gulf between environmental protection and the promotion of greater consumption in the energy sector is compromising the UK's environmental sustainability at all levels.

The privatisation of the electricity supply industry has brought many changes. In particular, the link between the coal and electricity industry, implicit in government policy until the early 1980s, was severed (this was one of the issues at the heart of the 1984 Miners' Strike). Instead, in a process begun under the former Central Electricity Generating Board, electricity generators sought to use cheaper fuels. Attempts were made to use a dirty heavy fuel oil called *orimulsion*, a fuel which essentially is a waste product produced from the extraction of the Venezuelan heavy oil deposits. This failed due to environmental objections. So the major change has been the switch from coal to gas – the so called *dash for gas*. Previously gas was seen as a high quality fuel that was more suited to domestic and industrial use. Then during the 1980s and 1990s, around the globe, the deregulation of the energy industry lifted the directive against burning gas for power generation. Generators were allowed to develop *combined-cycle gas turbine* (CCGT) plants that were more efficient, and so cheaper to run, than coal-fired plants (the differences between the two are described in the next chapter). However, it can be argued that this is the least efficient way to utilise our natural gas resources, and that using it to generate heat and power at the point of use could produce 30% more energy (see Box 11).

From the mid-80s onward, gas-fired plants have steadily taken over as the older coal-fired plants, many of which were built in the 1960s or earlier, were retired. This is in part due to the electricity trading scheme set up following privatisation which caused electricity prices to fall, making the older coal-fired plants uneconomic to run. But the overall capacity of the electricity supply industry has shrunk too. Under nationalisation, the supply industry kept up to 30% spare capacity. This provided additional power in the event that a number of power stations had to shut down. The privatised industry has shrunk this down to under 20%, close to the point considered necessary to ensure that the national grid can operate without the risk of blackouts [BBC, 2003c]. However, since the power blackouts in the USA, and more recently London, pressure is on the industry to maintain a higher level of spare capacity to guard against the potential for power shortages.

Energy Paper 68 forecasts how electricity generation will change over the next fifteen to twenty years (see figure 22). It projects future generating capacity, in giga-watts (GW) of electrical power, from various fuels. In 1990, power generation was dominated by coal, followed by gas and nuclear. Today it is gas that dominates power generation, but whilst the level of coal-fired generation will fall over the next fifteen years what will make the greatest difference is the closure of the UK's ageing nuclear power plants. Although the use of renewables is forecast to grow, most of the new generating capacity to replace the nuclear generation capacity will be gas-fired. Also many of the remaining oil-fired and older gas/mixed fuel (oil and gas) plants will close.

The quantities of gas now used for power generation, almost 50% of the UK's primary gas demand [DTI, 2002a], mean that gas reserves are being used at a much faster rate. Within the next few years the UK will be a net gas

Box 11. The "Dash for Gas"

Between 1990 and 2020 it is forecast that the capacity of plant generating electricity from natural gas will nearly treble [DTI, 2002c]. The amount of electricity generated from natural gas has increased from 0.6% of total generation in 1980, to 39.6% in 2002 [DTI, 2003a]. During the boom in building combined-cycle gas turbine (CCGT) plants in the late 1980s this trend became known as "the dash for gas".

What made gas so attractive was the economics. CCGT plants are more efficient than coal-fired plants. Of the coal fed into a power station, only 30% emerges as electricity. For a CCGT, 50% of the energy in the gas is converted to electricity. However, the CCGT is not the most efficient way to use natural gas.

About 7% of the electricity created by power stations is lost in distribution, so that 50% efficiency is really 46.5% by the time it gets to the consumer. If that gas were put into the ordinary distribution system and used directly in homes with condensing boilers, it could produce hot water and heat at 60% efficiency. In large commercial premises gas can be used in *combined heat and power* (CHP) plants, producing electricity and heat, at up to 65% efficiency. Recently developed systems can even produce heat and power in the home using micro-CHP systems [BBC, 2004h].

Given that a large part of everyday electricity demand is for space heating and hot water, it would make far more sense in energy terms to close down gas-fired plants and use the gas more efficiently at the point of energy use. In 2003, 29.7mtoe of natural gas [DTI 2004b], or 1.24EJ†, was burnt in gas-fired power stations each year. At 46.5% efficiency this creates 0.578EJ of usable electricity. If that gas were burnt at the point of use at 60% efficiency in condensing boilers it would yield 0.746EJ of heat and hot water – that's 0.168EJ, or 29%, more usable energy‡.

†To convert from mtoe to exa-joules (EJ), first multiply the mtoe figure by 0.041868. So, 29.8mtoe × 0.041868 = 1.25EJ.

‡1.25EJ of heat generating at 46.5% efficiency will produce [1.25EJ × 0.465 =] 0.581EJ/year. Burning 1.25EJ in domestic gas boilers at 60% would produce [1.25EJ × 0.60 =] 0.750EJ/year. That's [0.750 – 0.581 =] 0.169EJ, [0.169 ÷ 0.581 =] 29.0% more energy than using that gas in CCGT plants.

importer in order to supply the large demand for power generation. This will mean that gas will have to be piped across Europe, or shipped to the new LNG terminals being built in West Wales and the Thames Estuary [Earth Cymru, 2004], in turn adding to the insecurity of electricity supplies should

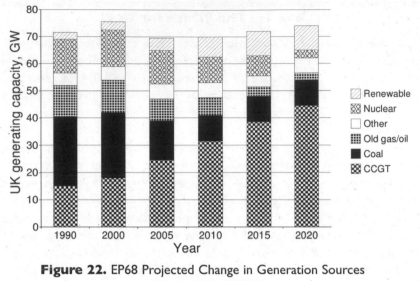

Figure 22. EP68 Projected Change in Generation Sources
[Source: DTI, 2002c]

these supplies of gas be interrupted.

Another consequence of using so much gas is that the UK has been able to meet its obligations under the Framework Convention on Climate Change. The UK makes up around 2% of the world's greenhouse gas emissions. In 2003, the biggest sources of carbon dioxide emissions were power stations (30%), industry (23%), transport (22%) and the domestic sector (16%). After reaching a low point of 150.3 million tonnes in 2002, these emissions have now begun an upward trend, and in 2003 rose to 152.3 million tonnes [DTI, 2004d]. The UK's greenhouse gas emissions are principally carbon dioxide (85%) and nitrous oxides (6%), and the remaining 9% is emitted as methane and volatile carbon compounds from agriculture and industry. In 2001, the UK's greenhouse gas emissions were 12% below the 1990 emissions target – meeting the UK's Kyoto Protocol target. The use of gas-fired plant has reduced the production of carbon, but as most of the nuclear plants will close over the next twenty years the carbon produced from the gas-fired plants they are replaced with will cause greenhouse gas emissions to rise more steeply.

Projections for the change in the UK's carbon emissions are given in Energy Paper 68. These are shown in figure 23, but in addition the most recent data for carbon dioxide from other DTI studies has been superimposed over the EP68 data to see how the model performs. The fact that it is higher than the project value is not a problem, provided that the trends between the two sets of data are consistent. The data is consistent with the lower estimate until

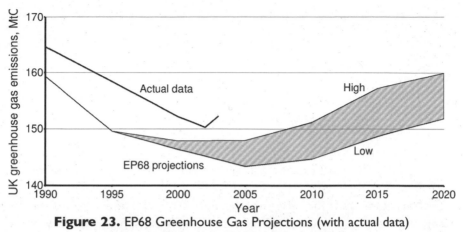

Figure 23. EP68 Greenhouse Gas Projections (with actual data)
[Source: Adapted from DTI, 2002c/DTI 2004f]

2002, but the sharp increase in 2003 would indicate the greenhouse gas emissions are likely to follow the higher trend, and the government has conceded that this rise might lead the UK to exceed the 2012 target for carbon emissions [Guardian, 2004o].

The UK's 1990 baseline for carbon emissions is about 200MtC/year (million tonnes expressed as carbon per year). The UK's Kyoto target, reducing emissions by 12.5% below 1990 levels by 2012, means that these need to be reduced to around 180MtC/year. Depending whether the UK's carbon emissions follow the high or low growth trend, which we won't know for certain until 2005 or 2006, then the Kyoto target will be physically exceeded before 2010 [DTI, 2004h].

The fact that we will physically exceed the Kyoto target is not a new issue. In fact, it is discussed in the UK's first report under the Framework Convention in 1994 [HMSO, 1994]. Therefore instead of seeing a real-terms cut in greenhouse gas emissions, other methods will be used to achieve an equivalent cut. As part of the recent energy white paper (white papers are a statement of future government policy), *Our Energy Future: Creating a Low Carbon Economy* [HMSO, 2003], the government has outlined various ways in which carbon dioxide emissions can be reduced or offset by other means.

As a means of preventing climate change becoming any worse the reduction target of the Framework Convention on Climate Change, implemented through the Kyoto Protocol, is insufficient. Consequently, and taking note of the advice from the Royal Commission on Environmental Pollution's recommendations [RCEP, 2000], the government has set a target of reducing carbon emissions by 65% of the 1990 baseline by 2050 – a total cut of over

100MtC/year. There is also an interim target of cutting between 5MtC/year and 25MtC/year by 2010. This target is made up of:

- Energy efficiency in households, 4 to 6MtC/year (25%);
- Energy efficiency in industry, commerce and public sector, 4 to 6MtC/year (25%);
- Transport, 2 to 4MtC/year (15%);
- Increasing renewable generation, 3 to 5MtC/year (20%);
- Offsetting carbon production by purchasing 2 to 4MtC/year (15%) of carbon credits through the EU's carbon trading scheme.

The target for transport does not include cutting the level of transport use in any way. Instead it focuses on continuing voluntary fuel efficiency agreements with car manufacturers and encouraging people to switch to biofuels – but switching to biofuels is not a simple as it sounds. As will be outlined in the chapter on renewable energy, this has serious implications for agriculture and land use.

The proposals for energy efficiency have very little detail associated with them, other than the outline of where savings could be made in an annex to the white paper [DTI, 2003b]. Of all sectors, it is the domestic sector that is highlighted as the area where the greatest savings might be achieved. However, the domestic sector is the smallest, consuming less energy than either the transport or the industrial/commercial sector (see figure 11). Proportionately reducing energy use by 1% in the transport sector saves more energy than reducing energy use 1% in the domestic sector. It is also comparatively easier to save energy in the transport sector through fuel switching or mode switching rather than trying to save energy through greater efficiency in the domestic sector. Therefore more energy could be saved if other sectors, particularly transport, were reformed rather than loading the focus for action on the domestic use of energy.

The white paper sets no particular target for the mix of energy sources in order to meet the carbon dioxide target. Instead, the energy industry will be encouraged to introduce measures, within the operation of the energy market, to meet the target. Renewable energy will be an option, but as the UK relies on centralised energy distribution networks this affects how alternative energy sources may be used. So from the point of view of the energy industry this leads to one dominant solution – more gas capacity. Many large energy users are installing *combined heat and power* (CHP) plants that burn gas to produce electrical energy and heat. This is more efficient than burning gas in centralised power plants, and some of the carbon savings predicted in the white paper are based upon a switch to CHP by large energy users. However, a switch to gas-fired CHP will build up a structural dependence within the commercial sector, as well as the power generation sector, on the availability of natural gas at low prices.

The white paper sets a target for renewable energy of 10% of electricity supply by 2010. Note the use of the term *electricity supply*, not *energy demand*. This has a very subtle meaning in terms of the overall energy system because of the low efficiency of electricity production. In 2003, the electricity generators produced 400TWh (tera-watt-hours), or 1.44EJ, of electrical energy. After deducting use of power by certain large energy users and the system losses, as is the process within the compilation of the official statistics, the *electricity supply* figure was 377TWh, or 1.36EJ (5% less). So the government's renewable energy target is 10% of this is figure – 0.136EJ. This is equivalent to just 1.9% of the UK's final energy supply in 2003, 7.13EJ, or only 1.3% if you base it on the UK's primary energy supply in 2003, 10.25EJ.

One interpretation of this poor target for renewable energy is that it takes advantage of the public's poor understanding of the energy sector. It is based on electricity, which only makes up 20% of the energy finally delivered to the UK economy, and so the target is bound to be incredibly small compared to the entire quantity of energy used. However, in the public's perception of energy use, 10% of the electricity supply sounds quite a lot. This is because electricity has more meaning in the public's perception than the abstract concept of the calorific value of petrol, diesel or natural gas.

Perhaps the most controversial element of the government's programme is *carbon trading*. Carbon trading, like past models of post-colonial economic development, requires that developing countries dis-appropriate their citizens of their carbon allowance in order to take part in the scheme. There is now a trans-national organisation, the International Emissions Trading Association (IETA), that co-ordinates work developing carbon trading systems between the industrialised states, multinational corporations and developing states. Already some organisations are gearing up to broker the finance as part of these deals, and the Dow Jones trading system now has an emissions trading board.

As more international carbon trading systems, such as the EU's carbon trading system, are agreed the price of carbon credits is rising. Up from £6/tonne in mid-2003 to between £8/tonne and £9/tonne in late 2003 [DJNW, 2003]. In Europe, the European Carbon Trading Scheme will broker carbon between the states and large energy users of the European Union. Here carbon prices were, in late 2004, £5.96 (€8.50) per tonne, although since unofficial trading from early 2003 prices have ranged between £3.51 (€5.00) per tonne and £9.40 (€13.40) per tonne [Pointcarbon, 2004]. The fact that carbon payments represent a substantial a flow of money between the rich and poor countries means that there are also a number of investment companies now offering carbon trading as part of portfolios to exploit the value of these payments. For example Eco Securities, a UK based investment firm, is brokering deals with carbon credit funded loans for construction projects in developing countries [EcoSec, 2002].

Box 12. Carbon Credits, Carbon Trading and Carbon Sinks

Inspired by the market leaders in energy trading in the mid-1990s, Enron, the energy sector is now a growth area within the global trade in commodities. Commerce likes to reduce things to simple commodities – that way something can be owned, traded, and a profit extracted. In turn, the market can define the terms upon which the commodity is created or used, usually to favour the largest producers.

Key amongst the new trading systems is the proposal for carbon trading. This involves the transfer of part of the per capita carbon allowance from countries with spare carbon to trade (usually poor countries), to countries that need a larger carbon allocation (usually rich countries), in return for money or other benefits. This has two effects. Firstly, it enshrines the reduction in carbon emissions to the spreadsheets of energy accountants. It is not necessary for the industrialised states who have become rich by burning fossil carbon to change their ways. Secondly, in developing nations, it removes the individual's right to use their carbon. If locked into long-term agreements to sell carbon, by international institutions like the World Bank, carbon trading could stifle the improvement of living standards in developing states.

As the commercial finance industry sees that there is money to be made from carbon, it's not just carbon emissions trading that is being offered on the market. There is a market emerging in *carbon sinks*. This involves planting large areas of land with trees or other crops that are intended to soak up carbon from the air. States, large corporations, and even individuals are then encouraged to buy the *carbon credits* created, and these provide the funding to pay for the scheme. But, as a means of arresting climate change, the concept is futile. For example, the forestry-based carbon credit schemes were condemned in 1999 by the Intergovernmental Panel on Climate Change because, as the Earth climate warms, the fixated carbon will be liberated with the first forest fire.

In the USA, some organisations are proposing to seed the oceans with nutrients such as iron and phosphates to increase plankton growth, and hence carbon fixation. However, like nutrient pollution in freshwater habitats, this in turn could have damaging consequences to the rest of the ocean ecosystem. Like other forms of carbon credit, ocean seeding would be funded by corporations purchasing carbon through the scheme, but research indicates that, like other carbon sinks, it may have an indeterminable effect [WHOI, 2000].

The problem with all these schemes is that they are run from the industrialised nations – and therefore a proportion of the profits will end up coming back to those states paying for the carbon credits. These deals are also seen by organisations such as the World Bank and International Monetary Fund as a potentially large source of foreign exchange, enabling developing states to repay their existing debts, or to fund new loan schemes. When carbon loans fund development projects, it's likely that the technical expertise will come from developed states, meaning that a proportion of the carbon payment will be returned to the countries paying it. This is even more likely to be the case where development aid is linked to or paid for as carbon credits. So although at one level carbon trading theoretically redistributes wealth to the developing world, in reality a proportion of these payments will be instantly returned via the banks and brokering agencies. Already some campaign groups associated with the anti-capitalist movement are looking at carbon trading as another means by which rich countries can exploit developing countries [CorpWatch, 2000].

The energy white paper proposes buying 2MtC/year to 4MtC/year of carbon credits. At current prices (averaging the range of prices in Europe given above, €9.20 or £6.46 per tonne) this will cost between £13 million and £26 million per year. However, if the price of carbon credits rise, and other measures do not produce the required savings in carbon emissions, then the total cost of reducing greenhouse gases by offsetting production with carbon credits would rise significantly. For example, as more states require carbon credits to meet their Kyoto targets then the price might rise as high as £60 per tonne. If the UK government had to buy 8MtC/year to 16MtC/year that would cost half a billion to a billion pounds. So as a policy solution, putting so much reliance on carbon credits is a financial gamble because it puts economic policy at the mercy of the international market in carbon trading. It's also worth noting that the white paper projects that by 2010 financial support for renewable energy, over and above the costs of carbon credits, will be running at £1 billion per year.

It is a fact that all projections are ultimately wrong. The purpose of any projection is to give an indication of possible outcomes within a certain level of probability, not to produce the correct answer. What matters is *how wrong* the projections are at the end of the period for which they are made. In the case of Energy Paper 68 and its projections, from the body of expertise available to the DTI it is likely (barring major financial meltdown or large rises in fuel prices) that the trends are roughly correct. This creates a problem for the government. It is committed to reduce carbon emissions, but it can't do that when all the data indicates that energy use is on the rise. So, in the white paper it has proposed carbon trading as a solution to these two incompatible trends.

A major criticism of the UK's energy policy, as outlined in the white paper, is that it's not a policy document about energy – it's a policy document about carbon dioxide emissions. It's at this point we can return to the issue of Peak Oil. If oil, one of the principle fossil fuels, is in imminent decline won't this solve a number of the problems? On the face of it, yes. If there is less oil, and we shift more towards gas use, then carbon dioxide emissions will reduce. But the carbon reduction policy is really just storing up problems for the longer-term. Natural gas production is projected to peak in the 2030s. If gas then becomes scarce, how will the UK's low carbon policy fare then?

More significantly, one of the intended consequences of carbon trading is to make fossil fuels more expensive in order to allow renewable sources to compete on price. As will be outlined later in the book, the major flaw in this argument is that there may not be enough land to provide a level of renewable energy sources equivalent to the amount of fossil fuels used today. In this case, the whole policy is doomed to fail because its approach is based on substituting for an ever-higher level of energy consumption, not controlling energy consumption within sustainable limits.

There are various ways in which we could project the effects of Peak Oil and Peak Gas on the UK's energy economy. But if we assume that the amount of the world's oil and gas resources that the UK uses today were to remain constant (in reality, unlikely after energy shortages begin to arise) then we can project the reduction in the UK's use of oil and gas as a proportion of the world's oil and gas production. On top of this we can add other energy sources. If we take the assumptions made in EP68 about coal, nuclear and renewable energy sources, we can project roughly all the UK's energy supply for the next century.

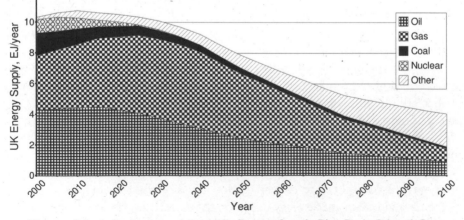

Figure 24. A Projection of the UK's Energy Supply Past Peak Oil and Gas

The projection of future oil and gas supplies, shown in figure 24, is based upon the assumptions of Laherrère and others [Laherrère, 2001; OU, 2003]. This assumes that the world total resource of conventional oil is around 2,000 billion barrels, with a further 750 billion barrels of non-conventional oil that will be produced over the latter half of this century. The world total resource of natural gas is taken as equivalent to 2,000 billion barrels of oil. Currently the UK uses 2.2% of the world oil supply, and 3.7% of the world gas supply. Figures for other energy sources are taken from EP68, but as this only provides data until 2020 it assumes that the trend in the EP68 data continues until the end of the century. For the sake of comparison the data is converted to exa-joules (EJ).

In 2003, the UK's energy consumption was 7.13EJ, although the UK's primary energy demand, which includes the systems and transformation losses, was 10.25EJ. What is significant is that the large increase in gas use does not create a large increase in the energy supply. This is because as the nuclear generating stations close, and as coal-fired stations are closed to reduce carbon emissions, all the increasing gas supply does is to leave the UK's energy supply at a stand-still. This illustrates the recent concern about the potential for power blackouts in the UK. Not just because the UK's power generation capacity is falling, but because the increasing reliance on gas makes the UK's energy supply insecure should the trans-European gas pipeline ever be severed [BBC, 2004i].

What figure 24 illustrates is an issue that many associated with the energy industry are concerned about. The combination of Peak Oil, and the failure to renew and expand the UK's electricity generation plant, is leading to a possible shortfall in energy. Clearly, if this projection were to be correct, in twenty years time the UK will have problems meeting its ever increasing demand for energy. The reason that the UK, or more recently China and India, has been able to use more oil and gas is because our use has been unconstrained by the productivity of energy resources. That will not be the case after the peak in oil, and then gas, production. After each of the peaks every country in the world will have to compete for a slice of an ever dwindling cake. But the fact is that after the peaks, when use is constrained, for one person in the word to have the same, or greater, supply of oil and gas another person must have less.

In reality the drop in the availability of energy resources in twenty years will not happen. It will not happen because in the face of such a crisis the government is likely to adopt other energy sources, irrespective of the costs associated with their use. The clear frontrunner here is nuclear power, although new technologies based upon coal are also likely to be used.

However there is potential solution, related to the Peak Oil issue, that few official bodies have seriously considered. It's not been considered because within the current economic framework, and centralised energy systems, it's just not conscionable. *What if we just stopped using so much energy?* That does-n't mean switching off, but rather a bottom up re-think of how we source and use the energy we need. The problem with such a solution, especially for the business-centred DTI, is that it represents not business as usual, but *business as unusual*. It means redefining what the energy industry is there to do. The greatest problem for any government is that cheap, abundant fossil fuel ener-gy has infiltrated the entire fabric of society. But even organisations such as the Organisation for Economic Co-operation and Development (OECD) are now saying that the era of cheap energy is at an end [Guardian, 2004p]. We have to think of new energy solutions based upon a different set of assump-tions about cost, efficiency and environmental impact. Therefore any modifi-cation to the basis of the energy industry must affect the public, business and social systems of the entire state.

When thinking about the gap between energy supply and demand we shouldn't just concentrate on how we might produce more energy. Greater efficiency in how we use and convert fuels will allow the available energy to go further. Likewise, if alternative energy sources were expanded, how much could they produce? To understand more we have to look at conventional and renewable energy sources in detail. We need to know what they can produce, and what problems may result from their use. We also need to look at how energy could be saved.

Over the next three chapters we'll look at all these issues. First, conven-tional and nuclear energy sources. Secondly, low carbon sources, which are strictly not renewable but do not produce large amounts of carbon dioxide. Thirdly, renewable energy sources. This leads us, finally, to the most difficult question of all, which is presented over the last two chapters of the book – *do we need to cut our current use of energy in order to match the levels of energy that may be available to us in the future, and if so, by how much?*

6. Conventional and Nuclear Energy

In this chapter conventional includes all forms of fossil fuel energy, nuclear energy and burning waste. All these sources of energy, although very different, work in fundamentally the same way. Most of the fuel used by industrial society provides large quantities of heat. This either heats air spaces or water. In UK homes, about a half of energy consumption is used for heating the air and a quarter is for heating water [DTI 2002b].

The largest use of heat is for electricity generation. The heat boils water to create high pressure steam, then the steam pressure turns an apparatus that delivers the electrical energy. All these power generation systems form what are called *heat engines* (see Box 13).

Although society uses many types of fuel that contain energy – such as coal, petroleum, gas, nuclear and even industrial and household waste – as a society we only use a very few forms of energy. Mostly heat, electricity, and kinetic energy (moving things). Energy conversion processes produce heat as a by-product, reducing the efficiency of conversion. Consequently, we never get as much energy out of a process as we put into it. When examining any energy system, conventional or otherwise, it's important to look at the conversion process(es) involved and the efficiency with which they produce usable energy.

Gas- or steam-driven piston engines, also called *reciprocating engines*, are a rather wasteful form of power generation. The first steam engines were developed by Thomas Newcomen in 1705. In 1760, James Watt redesigned the steam engine to use high pressure steam. This not only increased the power output, but also the efficiency (see Box 13), which allowed them to drive the large machines that created the Industrial Revolution.

Many steam engines operate at an efficiency, from the energy content of the coal transformed into motion, of around 4% to 8%. The great advance in using steam came at the end of the Nineteenth Century when Charles Parsons invented the *steam turbine*. Rather than expanding the steam in a chamber containing a piston the steam expands continuously through a series of rotor blades. This means that on one side of the turbine is a high pressure, on the other side is a low pressure, and as it expands through each set of rotors the steam transfers kinetic energy to the blades, turning the shaft that they are fixed to. To work efficiently the rotor blades must turn at half the speed of the steam jet, which can mean the turbine rotates at a few thousand revolutions per minute. The latest generation of turbo-fan jet engines, which are gas turbines developed from the principle of the steam turbine, contain

Box 13. Heat Engines and Conversion Efficiency

The theory of heat engines was developed during the Nineteenth Century alongside the development of the steam engine. In the early Nineteenth Century a French engineer, Sadi Carnot, described the principles of heat engines theoretically and worked out that to get a highly efficient heat engine you had to engineer the greatest possible difference between the temperature of the heat input (T_0) to the engine and the heat output (T_1) from it. So the hotter the steam going in, and the colder the output, the more energy you can extract from the system. Note also that Carnot engines are idealised because they do not have any energy losses. In reality heat engines are not as efficient because of the loss of heat, steam leaks, and the friction generated between the various moving parts.

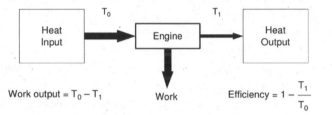

Work output = $T_0 - T_1$ Work   Efficiency = $1 - \dfrac{T_1}{T_0}$

Figure 25. Carnot's Heat Engine

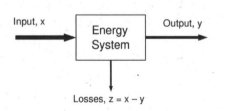

Efficiency = y / x

Input, x   Energy System   Output, y

Losses, z = x − y

Figure 26. Measuring Efficiency

Heat engines and other energy conversion processes are often quite complex to analyse. Therefore the energy movement in a complex system is usually assessed in terms of the *efficiency of conversion*. Rather than model each part of the transformation process, an efficiency figure is calculated as a ratio of the energy output to the energy input. Using a general measure of efficiency allows us to simply assess the inputs to a transformation process, and then calculate the useful energy or work produced, and the loss of energy from the whole system. Today most of the energy systems we use still operate at less than 50% efficiency, and only a very few operate at efficiencies higher than 70%.

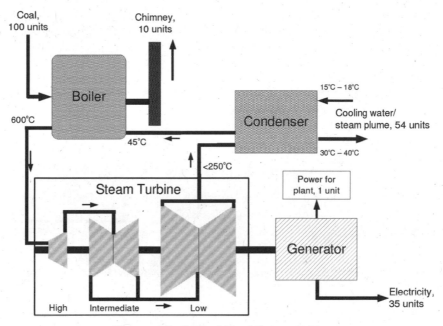

Figure 27. A Coal-Fired Power Station

turbines operating at over ten thousand revolutions per minute.

To increase the efficiency of operation modern steam-raising power plants use groups of turbines that operate at different pressures – high, *intermediate* and *low* (see figure 27). The temperature of the steam going into the high pressure turbine is 600°C, and it emerges from the low pressure turbine at around 250°C. This cooled steam must be then turned back into water by passing it though a condenser where the steam is cooled with river water or air. It's the condenser that produces the familiar steam plume that rises from power plants. Using the latest steam turbines the proportion of energy that can be extracted from the fuel rises to 30% or 35%.

Analysing how a coal-fired plant works (most nuclear plants use the same system too), if you put 100 units of coal energy in the boiler, 10 units are lost via the chimney, 54 units are lost from the condenser as a steam plume and warm water, and 1 unit is consumed by the power station itself. This leaves 35 units to be exported as electrical energy. So, for coal-fired plants, of the original 100 units of energy in the coal 64 are lost to the environment as waste heat.

The older oil- and gas-fired plants are similar in operation to coal fired plants, and have similar efficiency, but combined-cycle gas turbine (CCGTs) plants are more complex (see figure 28). Instead of one generating cycle there are two. The first cycle operates from the gas pressure of the burning gas, using

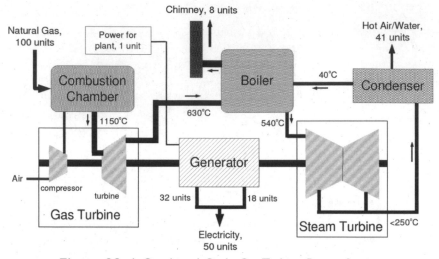

Figure 28. A Combined-Cycle Gas Turbine Power Station

a gas turbine to convert the gas pressure into electrical energy. The hot gas leaving the gas turbine passes to a boiler where it heats water to produce steam for the second generation cycle. The steam passes to a steam turbine, and then on to a condenser. The combined efficiency of the two generating cycles from the single energy source is around 50%. If you put 100 units of gas energy into the plant you get 32 units of electrical energy from the gas turbine and 18 units from the stream turbine. Of the remaining energy, 1 is used to run the plant, 8 are lost from the chimney and 41 are lost from the condenser.

As the emission of pollution is proportionate to the efficiency of combustion, this also means that CCGTs emit less carbon dioxide and other pollutants per unit of fuel burnt than coal- or oil-fired plants. Given this higher efficiency coal-fired technology would seem to be a obsolete, but there is another option for creating energy from coal that is largely unused because of the cheap cost of gas. If coal is gasified — turned into a gas by heating it to high temperatures — the gas can be cleaned and burnt in a combined-cycle gas turbine at much higher efficiencies than a conventional coal-fired plant. This type of plant is called an *Integrated Gasification Combined-Cycle* (IGCC) plant. Apart from the fact that the emissions are cleaner than an ordinary coal-fired power station, there is greater scope to control the higher carbon dioxide emissions associated with coal-fired plant. The plant itself produces less carbon per unit of power generated, but it would be possible to install carbon sequestration technology to reduce emissions still further. This uses compounds to soak up the carbon dioxide producing a waste stream that, it is proposed, could be injected back into old oil wells to isolate it from the environment. However, all this extra processing means that, from fuel to power, IGCC is about 45% efficient. Slightly lower than a CCGT, but higher than a standard coal-fired plant, and with carbon sequestration it would produce far less carbon than a comparable gas-fired plant.

The only way that the efficiency of any thermal energy plant could be improved further, be they coal-fired, CCGT or IGCC, is by finding a use for the large quantities of heat they release to the environment. It might be possible to co-locate a large user of heat energy, such as a food processing plant or horticultural greenhouses, next to a power station. This plant could utilise the waste heat from the plant, rather than feeding it into a condenser, which would significantly improve the overall efficiency. The more expensive option is to install a district heating system. This is the equivalent of having a central heating system that serves an entire town or neighbourhood, supplying each house with heat energy for space and hot water heating. However, unlike Europe where district heating is more widely used, in the UK we tend to put our power stations far outside urban areas – making district heating impractical. If small-scale district heating systems could be developed alongside a combined heat and power plant that served a neighbourhood or a commercial complex this could increase the efficiency of energy use to 60% or 65%.

Another problem is that both coal- and gas-fired plants produce large quantities of pollution. Coal fired plants produce a lot of pulverised fuel ash (PFA). A proportion of this is recycled to make building blocks, and it is used as a fill material in the construction industry. However a large amount is landfilled back into old mineral workings. Where there is a high water table, such as in old gravel pits, this can give rise to groundwater contamination. The heavy metals and other compounds in the ash, such as boron, are flushed out by groundwater movement and can enter local streams and rivers.

All combustion plants produce a variety of *combustion gases*. Combustion produces nitrogen oxides, which reduce air quality, and can cause ill health amongst those susceptible to air pollution. Nitrogen oxides also cause *acid deposition* (or acid rain) as the nitrate mist reacts to form nitric acid. Coal-fired plants have the additional problem of emitting large quantities of sulphur compounds. This problem has reduced since the 1980s, when the UKs sulphur emissions created acid rain that damaged forests in the UK and Europe, because many power stations now burn low sulphur coal. Also new European regulations, to be implemented in 2008, would require that sulphur scrubbers are fitted to all coal-fired plants to reduce sulphur emissions still further. This directive has the potential to close many of the UK's coal-fired power stations because retrofitting scrubbers would not be very economic because most of the UK's coal-fired plants are more than 30 years old.

IGCC plants, because they are run as small chemical plants, have the ability to recover the pollutants, such as sulphur, and recycle them as raw chemicals for use in industry. This would off-set the need to produce them using other energy resources. IGCC plants could also lower the nitrogen content of the air used in the plant to reduce nitrogen oxide emissions, but an IGCC plant would still emit pollutants.

Efficiency is often key not only to energy use, but to associated issues such as pollution or waste generation. The UK's coal-fired power stations, most of which are more than 30 years old, convert coal to electrical energy at about 30% efficiency. So to produce 1 unit of electrical energy you must burn 3.33 energy units of coal (at 30% efficiency, producing 1 unit of energy takes [100% ÷ 30% =] 3.33 units of energy). CCGT power stations operate at around 50% efficiency. Therefore you only need burn 2 units of gas energy to get 1 unit of electrical energy ([100% ÷ 50% =] 2 units of energy). This lower level of energy wastage, to achieve the same power output, also means that pollution levels decrease in the same proportion. As a general rule, pollution and waste always follow energy use, and are produced in proportion to the efficiency of the process.

Another option, rather than burning primary fuels, is to burn waste. For some time local authorities, and especially industry, has been burning waste materials as a source of heat or power. For example, cement kilns have been burning waste solvents, and more recently car tyres, to reduce the use of coal used in the process. One of the motivations for this is that instead of an organisation having to buy expensive fuel for their process, waste disposal companies will pay these energy users to burn their waste.

The problem with most types of waste incineration is that the waste material is often contaminated, and these contaminants will be passed through the process and into the gases emitted from the plant. For example, car tyres contain cadmium that can increase the toxicity of local air pollution. The less volatile contaminants from waste incineration usually end up in the ash stream, and this can create a serious pollution hazard. In particular, the ash from the incinerator's pollution control system (called a *scrubber* or *precipitator*) which concentrates the pollutants cleaned from the flue gases.

The greatest problems with waste incineration occur when the waste stream is highly *heterogeneous*. That is, it is made up of many different waste types. Using waste as a source of heat is easier when the waste stream is *homogeneous* – when it is made up of exactly the same type of waste. For example, waste paper and wood are used as a heat source, usually as a way of reducing the waste produced and the energy used by the factory that is creating these materials. Heterogeneous waste streams, because they often come from a number of different sources, often contain a wider range of pollutants. The benefit of a homogeneous waste stream is that it is far easier to track the pollution content of the waste because it usually originates from a more restricted number of sources. Consequently it is possible design a more suitable pollution abatement solution for homogeneous wastes to prevent high concentrations of harmful pollutants being emitted to the air. However, any waste combustion process, the more you scrub the combustion gases clean the more toxic the waste material produced from the scrubber plant. It's the *Law on the Conservation of Matter and Energy* – what goes in must come out, somewhere. The better the pollution control plant is at cleaning pollution from the flue gases, the more toxic the scrubber residues that will be produced.

One of the options the UK government have been promoting for some time is burning, or incinerating, household and commercial waste to create electrical energy. These plants use a specially designed combustion chamber, with a bed of sand (a fluidised bed incinerator) or a mechanical grate (an agitated grate incinerator), that continuously burns the waste fed in. The heat this produces raises steam which, like conventional power plants, turns turbines to produce power. The waste heat from the plant might also be supplied to a nearby energy user, or to a district heating system.

Figure 29 shows the layout of an average municipal solid waste (MSW) incinerator. MSW incinerators take a highly heterogeneous waste stream. Rather than a segregated collection of waste, which enables a higher quality of materials recycling, incinerators usually accept waste collected in bulk. This is mechanically pre-sorted to remove some of the recyclable content. One of the main purposes of pre-sorting is not to enable recycling, but to remove non-combustible material, like builders' waste, or objects such as pressurised gas cylinders that might damage the incinerator.

The DTI claim that incineration is a renewable technology, but no objec-

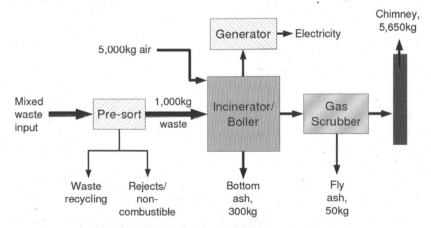

Figure 29. A Municipal Solid Waste Incinerator

Box 14. Incineration, Recycling and Lifecycle Assessment

To get a better appreciation of the efficiency of a waste incinerator we need to look at the whole energy cycle. This type of approach, where you assess an activity or commodity over the process of creation and use, is called *life-cycle assessment*, or LCA. LCA has become an important analytical tool within waste management and energy economics. When assessing how society uses energy, and whether those uses are sustainable, a much higher level of detail is required. LCA is the best method to provide this detailed information.

Table 3. Energy and the Incineration or Recycling of Waste
[Source data: White, 1994]

Waste type	% mixed waste by weight (%)	Calorific value, GJ/te	Energy used to manufacture, GJ/te	Energy used to recycle, GJ/te	Net incineration energy, GJ/te	Net recycling energy, GJ/te
Paper	34	10.5	22.7	14.4	2.6	5.4
Glass	10	0	9.6	5.8	0	2.5
Ferrous metal	5	0	33.5	20	0	8.8
Non-ferrous metal	1	0	171	15.6	0	101
Dense plastic	4	28	91	7.6	7	54.2
Plastic film	6	25	88.5	25.4	6.3	41
Other	40	4	0	0	1	0
Mixed waste	100	7.8	21	8.5	1.9	8.2

Incineration value calculated as the calorific value x 25% generation efficiency.
Recycling value calculated as [manufacturing energy − recycling energy] x 65% reclamation.

The net energy value of incinerating waste materials is in almost every case less than the energy reclaimed by recycling − by a factor of two or four more. For example, if waste paper, calorific value 10.5GJ/te (giga-Joules per tonne), is burnt to generate power at 25% efficiency it produces 2.6GJ/tonne of electrical energy. Producing virgin paper takes 22.7GJ/te of energy, but producing recycled paper takes 14.4GJ/te of energy. If we recycle 65% of all the paper in the waste stream each 1 tonne of paper saves 5.4GJ of energy[†] − twice as much energy as is produced by burning it to produce electricity.

The most interesting waste stream is non-ferrous metal – mostly made up from aluminium. This cannot be mechanically recovered as it is not magnetic, and is usually destroyed in the incinerator. A 150,000 tonne per year incinerator will destroy roughly 1,500 tonnes of non-ferrous metal per year. Assuming that this material must be re-manufactured, this represents 256.5TJ of energy – more than three-quarters of the electrical energy production of the incinerator‡.

† Incineration at 25% generation efficiency produces [10.5GJ/te x 0.25 =] 2.6GJ/te of energy from waste paper. Recycling 65% of the paper in the waste stream saves [(22.7GJ/te – 14.4GJ/te) x 0.65 =] 5.4GJ/te of energy.

‡ Non-ferrous metal is 1% of the waste stream, so at 150,000 tonnes per year an incinerator burns [150,000 x 0.01 =] 1,500 tonnes. The energy taken to manufacture this non-ferrous metal is [1,500 te/year x 171GJ/te =] 256,500GJ/year. Burning 150,000 tonnes of waste, calorific value of 7.8GJ/te at 25% efficiency, will generate [150,000 te/year x 7.8GJ/te x 0.25 =] 292,500GJ/year. So [256,500 ÷ 292,500] 88% of the incinerator's power output is wasted by destroying the non-ferrous metal within the waste stream.

tive analysis could classify it as such because it is far less energy efficient than options like recycling (see Box 14). In fact, it is arguable that incineration produces more carbon than conventional fossil fuels because most of the waste material is manufactured or processed with fossil fuels.

More significantly, if we look at the whole energy cycle, incineration *wastes* a large amount of energy. Individual waste materials have their own individual calorific values – the quantity of heat produced by burning them in the incinerator (usually measured in giga-Joules per tonne of waste, or GJ/te). So the total calorific value of mixed waste is dependent on the composition of the waste it contains (see Box 14).

Each individual waste material also has its own energy value for production, and for recycling. Whilst recycling saves energy because the recycling value is usually less than the production value, incineration destroys these resources and so they must be re-manufactured, requiring that the energy of production is expended again. Consequently the efficiency of incineration is low not just because of the low level of efficiency of steam turbines, but because the materials burnt must be re-manufactured and this takes more energy than recycling them.

The types of waste that can be most easily recycled, such as paper, card and plastics, are also the waste types with the highest calorific values. This creates a conflict between the need to maintain the energy value of the waste that must be fed to the incinerator, to produce power, and the recycling of waste. Although the UK government and many local authorities regard incineration as recycling, if you look at the total impact of burning potentially recyclable materials this isn't the case. Across the whole waste stream incineration produces 4.3 times less energy than recycling would save. The commercial and municipal waste burnt in MSW incinerators is also one of the most heterogeneous waste streams, and hence liable to contamination with anything from spent household chemicals to toxic batteries and commercial radioactive materials.

It it also important to note that, unlike burning fossil fuels, incineration creates two releases of carbon dioxide and other greenhouse gases. One when the material is incinerated, and another when it must be re-manufactured to replace the resources lost in the incinerator. Recycling too produces two releases of greenhouse gases, but the total release much is less than the emission created by re-manufacturing the materials destroyed in an incinerator.

The other large-scale conventional (conventional, in the sense of a tried and tested system) source of energy we have today is nuclear power. Recently, with the arguments over carbon dioxide and climate change, the nuclear industry have argued that only nuclear power can address the coming gap in energy supply. Consequently the nuclear industry have put a lot of effort into lobbying governments across the globe to develop more nuclear power stations [Guardian, 2004n]. However, this argument masks the fact that nuclear power does produce carbon dioxide, but as part of the whole nuclear fuel cycle rather than from the power station itself.

An early study from the UK put the greenhouse gas emissions from British nuclear power stations at 3.9kgC/GJ (kilos, expressed as carbon, per giga-Joule of energy produced), about one-tenth the emission of modern CCGT plants [ETSU, 1990]. A more recent German study puts this figure at 9.4kgC/GJ for German nuclear reactors, but cites a range of other figures quoted in international studies between 8.3kgC/GJ and 17kgC/GJ [Öko-Institut, 1997]. So nuclear power not only produces greenhouse gases, but the level of those emissions are only an order of magnitude (one-tenth) of a gas-fired plant, not a minute fraction of them.

Like oil and gas, the uranium that is used to produce nuclear fuel is a finite resource. The availability of uranium is graded according to the cost of producing it from uranium-bearing ores. There are just over 1 million tonnes that can be recovered for less than $40 per tonne, nearly 1.4 million tonnes for between $40 and $80 per tonne, and just over 700,000 tonnes for between $80 and $130 per tonne [IEA, 2001]. In total, in terms of all

Box 15. How Nuclear Reactors Produce Energy

Coal, gas, and household waste produce energy by breaking the chemical bonds held together by the electromagnetic force. The strong nuclear force that holds together atomic nuclei is far stronger than the electromagnetic force, and so when we fission (break up) atoms it releases more energy. This is why it takes only a small amount of nuclear fuel to produce a large amount of energy. Even so, the heat energy created by the splitting of atoms can only heat water, in the same way that the fuel in a conventional power station heats water to produce steam. This means the production of electrical energy from nuclear sources is still just 30% to 35% efficient.

Standard thermal reactors split atoms of the isotope uranium-235 (^{235}U) to release energy. But the most common uranium isotope is uranium-238 (^{238}U) – ^{235}U only makes up 1 in 140 of all uranium atoms because it decays faster. So uranium is enriched to raise the levels of ^{235}U from the natural 0.7% to around 3.5%. This provides enough ^{235}U atoms to allow a chain reaction to be established within the nuclear fuel.

Each 1 kilo of ^{235}U, when it undergoes fission, can create 82TJ of energy – although an additional 50% will be produced by other nuclear reactions as some ^{238}U is converted to plutonium which is then fissioned to produce more energy. At an enrichment level of 3.5%, each one tonne of nuclear fuel contains 35 kilos of ^{235}U and 965 kilos of ^{238}U. When one tonne of nuclear fuel is used and removed from the reactor after three years, only 7 kilos of ^{235}U remain, along with 940 kilos of ^{238}U. The rest of the one tonne mass is made up of 9 kilos of plutonium, and 44 kilos of other radioactive elements.

The fact that only a small part of the uranium, the ^{235}U, is used as fuel means that there are large quantities of unused ^{238}U left lying in stockpiles at fuel fabrication plants around the world. This unused uranium is so worthless that is has begun to replace more expensive metals, such as lead, in uses such as the ballast weights of ships and aircraft. Depleted uranium has also become an important component in a number of modern armaments systems.

More significantly, as only 0.7% of the uranium resource is used to produce power, only 0.7% of the energy held in uranium resource is useful. The reason that the nuclear industry talk of hundreds of years worth of nuclear power is that they include in their figures the use of the ^{238}U in fast breeder reactors. These convert the ^{238}U into plutonium, and then fission the plutonium to produce energy. However, no commercial fast breeder reactor has yet been produced because of the safety issues associated with its operation. In addition, the large-scale use of fast breeder reactors would require a number of nuclear fuel reprocessing plants, like Sellafield, to be constructed. Consequently, if we were to develop a fast-breeder fuel cycle it would required a massive re-engineering of the nuclear industry.

extractable reserves irrespective of costs, it is estimated that there are 3.4 million tonnes of uranium reserves, although other sources put this at 2 million tonnes [EC, 2001].

Nuclear electricity currently makes up 17% of the world's electricity production, supplied by around 440 nuclear reactors in over 30 countries. However this is only 6.1% of the world's total energy use [BP, 2004]. Some countries, such as France and Lithuania, supply over 70% of their electricity from nuclear power. In 2003, the UK supplied 23% of electricity production [DTI, 2004a] from nuclear power, which represents 8.6% of the UK's primary energy supply [DTI, 2004b].

The problem with the existing thermal nuclear power stations is that we only use 0.7% of the uranium resource to produce usable energy. The rest is wasted (see Box 15). This creates a problem of perception because some analysts will use the 0.7% figure when expressing what is available from nuclear energy, whilst others assume that the entire resource is usable (which, using current technology, it is not). At the current level of utilisation, there are 100 years of nuclear fuel left to use, but if the world significantly expands nuclear power this 100 years figure will rapidly diminish.

This confusion over how much energy nuclear power can provide has led to confusion between political policy makers and energy analysts. For example, the British Prime Minister's think tank, the Downing Street Performance and Innovation Unit (PIU), described uranium in their review [PIU, 2002] of potential power sources for the recent energy white paper as:

> "...plentiful, easy and cheap to store, and likely to remain cheap. This means that nuclear power is essentially an indigenous form of energy."

The use of the word plentiful sounds as if the UK has a lot of uranium. We do, but it's nearly all depleted uranium left over from the production and reprocessing of nuclear fuel, and it is useless in thermal reactors. In fact, uranium is only abundant in the USA, Canada, Australia, South Africa, Brazil, Namibia, Uzbekistan, Kazakhstan and Russia. Plus the fact that the PIU report doesn't give a time-scale over which this abundant energy supply will produce usable energy.

Others take a different view of the availability of uranium. As energy demand grows, and if states begin to take the nuclear option to reduce greenhouse gas emissions, then the number of nuclear plants must grow significantly. Research published by the Organisation for Economic Co-operation and Development [OECD, 1999b] states that, for nuclear power to keep its current proportion of electricity production and to replace fossil fuels for energy production, then,

> "...the number of nuclear power plants would need to increase 30 times, leading to a total of 12,000 plants... Known uranium reserves would then last only for about a decade unless the reactors presently employed are replaced by breeder reactors."

So the nuclear option would not in itself be a solution with current nuclear technology. The reserves of uranium would run out equally as quickly as oil and gas. Instead we would have to develop fast breeder reactors. When thermal reactors fission uranium-235 (^{235}U) atoms the radiation given off turns some of the uranium-238 (^{238}U) into plutonium-239 (^{239}Pu). Of all the energy produced by thermal reactors only two thirds comes from ^{235}U. Much of the rest comes from the fissioning of the ^{239}Pu that is bred from the ^{238}U. The idea of a fast breeder reactor is that a central core of made up of ^{239}Pu and ^{238}U that undergoes fission to produce energy. At the same time a blanket of ^{238}U around the core is bred to produce more plutonium, which creates the fuel for future reactors when the fuel is reprocessed. This also means that the use of reprocessing plants, like Sellafield in the UK, or La Hague in France, are an implicit part of the fast breeder system.

The problem with this approach is the design limitations of the fast breeder reactor system. There are various experimental designs for fast breeders in operation, but most of these projects have been held-up by technical and safety difficulties. The UK's fast reactor programme was abandoned in the late 1980s when the government opted for the American designed *pressurised water reactor* (PWR) – which is a thermal reactor system. Like some PWRs, fast breeder reactors use a mixed oxide fuel made with oxide compounds of fissile ^{239}Pu and ^{238}U. In theory oxide fuel is safer than metal fuel because it is more stable at high temperatures, and it can't burn or react strongly with water (because it's already an oxide).

Thermal reactors have to include materials such as graphite or water (called the moderator) to slow down the neutron radiation given off by the fissioning of ^{235}U atoms. This allows a chain reaction to take place. Without a moderator the neutrons travel too fast to create fission in the ^{235}U atoms. This means the core of a thermal reactor is very big because of the space the moderator takes up. For example a pressurised water reactor, such as Sizewell B in Suffolk, has a reactor core that is 3.6 metres high and 3.4 metres in diameter. Fast breeder reactors use the fast neutrons directly, and so don't need a moderator. This makes the core much smaller – perhaps 1.5 metres high and 2.5m in diameter – even though the power output would be the same. This means that the *power density* of the core is more than four times higher.

This creates a major engineering problem. If there was an interruption in the cooling of the core of a thermal reactor it could melt down and explode in a hour. A fast breeder reactor could melt down and explode in a minute.

Apart from the safety issues related to the power density of a fast breeder reactor's core, the other issue is the potential hazards of a catastrophic failure. The use of large quantities of plutonium in fast breeders make the fuel a far greater radiological hazard if there is an accident. The use of fissile plutonium is also a hazard in terms of nuclear weapons proliferation. Currently plutonium is closely controlled by the International Atomic Energy Agency, as is fissile ^{235}U, under a system called *safeguards*. With more and more plutonium being moved around in nuclear fuel there is an increased risk of plutonium going missing.

The fuel availability, safety and efficiency issues aside, there is one problem with nuclear power that irrespective of the reactor design mitigates against its use – waste disposal. The current generation of 440 nuclear reactors has created 150,000 tonnes of highly radioactive radioactive waste [OECD, 1999b]. This type of waste needs constant cooling and monitoring for up to fifty years after it leaves the reactor as the decay of the most highly radioactive materials produces a lot of heat (and like a nuclear reactor, the waste could melt down and produce a large emission of radioactivity). Even after this initial stage the spent fuel must be kept in isolation for over one hundred thousand years to allow the intensely radioactive elements within the fuel to decay. The problem is that human society's oldest monuments are only a few thousand years old, and these are just heaps of stone. Creating a reliable structure to isolate radioactive waste for a hundred thousand years is very difficult.

Most solutions to the disposal of radioactive waste involve burying the waste deep underground. The difficulty for the designers of these deep repositories is that as radioactive materials decay they turn into different chemical elements, and these elements have different physical properties. The greatest problems occur when radioactive decay causes a change of phase. For example radium, a solid, decays into radon, which is a gas. So any potential storage system would have to withstand large internal gas pressures, otherwise it would rupture and release its radioactive load. For this reason the types of geological formation proposed for storing material have either very low levels of groundwater movement, such as clay, or the rock is naturally dry, such as salt domes or desert rocks.

As yet there is no large scale facility in operation for the disposal of spent nuclear fuel in perpetuity anywhere in the world, although there are a number of facilities around the world disposing of low and intermediate level wastes in shallow repositories. The USA is attempting to commission a facility at Yucca Mountain, in Nevada, against strong legal and scientific challenges. In reality, there is no certain storage method for the 150,000 tonnes of highly active spent fuel produced to date. If we take the projection given earlier, that 12,000 thermal reactors would be required by 2050, and scale up the level of spent fuel production, this would result in 4,500,000 tonnes of spent fuel by the end of this century.

As long ago as 1976, the Royal Commission on Environmental Pollution stated [RCEP, 1976],

> "It would be morally wrong to commit future generations to the consequences of fission power on a massive scale unless it had been demonstrated beyond reasonable doubt that at least one method exists for the safe isolation of these wastes for the indefinite future.".

On that basis, thermal- or fast-fission-based nuclear power is, irrespective of reactor technology, a non-starter.

The eventual saviour of the nuclear industry is, for many, the development of *nuclear fusion*. The heavy isotopes of hydrogen, deuterium and tritium, are heated to 10 million degrees to form a plasma. At this temperature the atomic nuclei fuse together, in the same process that takes place at the heart of the Sun, to form helium and an excess of heat energy. In order to hold the high temperature plasma a doughnut-shaped array of electromagnets confine the electrically charged atomic nuclei using magnetic forces. The excess heat produced, theoretically up to 350TJ per kilo of hydrogen isotopes, would make the plasma self-sustaining, at which point energy could be tapped to provide heat, and by a process of electricity generation, power. However, to date self-sustaining nuclear reaction has only been maintained for a fraction of a second, and a viable fusion power plant may be fifty to one hundred years away. Consequently, fusion power is well beyond the time-scale of Peak Oil and Peak Gas.

Although fusion is usually portrayed as a clean form of nuclear energy, it is not – but compared to fission you could probably call it *grubby*. Although it would produce inert helium gas as a by-product it would make the reactor vessel and the equipment around it radioactive, albeit to an intermediate level of activity rather than a high level. Once built, the reactor would need constant maintenance to keep the inside of the plasma chamber in good order, so a constant stream of radioactive waste would still be produced. All this material would have to be stored for fifty to one hundred years to allow this activity to decay before it could be disposed of as low-level radioactive waste.

Box 16. How Electricity is Generated

Electricity can be created in two ways – by *induction* using magnets, and by chemical reactions.

The first electric currents were created chemically, by reacting metals with acids. This releases a constant stream of electrons that creates a current flowing in one direction – a *direct current* (DC). Today we use a variety of battery technologies but the principle is the same – an acid or alkali electrolyte reacts with a metal to produce a source of electrons that creates electric current. As well as one-way chemical reactions, where the battery degrades and has to be disposed of, some types of battery allow the process to be reversed and the battery can be recharged.

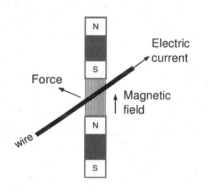

Figure 30. Electrical Induction

Most of our electricity is created by magnetic induction in a generator. If a wire that is part of an electric circuit is passed through the gap between two magnets the magnetic force interacts with the wire to induces an electric current (I) to flow. The magnetic field actually resists the passage of the wire, and so it is the energy within the force (F) pushing the wire through the gap that is transformed into electrical current, not the magnetic force from the magnets. Power generators use coils of wire in order to pass a long length of wire through the magnetic field and generate large electrical currents at high voltages. The kinetic energy from the motor, steam turbine, or water wheel driving the generator is usually transformed into electrical current at a fairly high efficiency – 60% or higher.

Some of the early power systems used direct current, but apart from the safety problems, the major problem was transforming the voltage from one level to another. For this reason most power systems use *alternating current* (AC). AC is produced as the circular coils of wire pass through magnetic fields with alternating polarities inside the generator (or, more correctly, alternator). The useful feature of AC is that you can wind wires into coils to create a magnetic field – the reverse process by which the current was created. When another coil is placed next to the first the magnetic field will induce a current in the second coil. Depending upon the ratio of the number of turns in the first and second coils the magnetic field will then step the voltage in the second coil up or down relative to the first. This arrangement is called a *transformer*, and it's by using transformers that it is possible to create national power grids.

Be it coal, gas, or nuclear, a large part of the UK's primary energy supply is used to generate electricity. Even so, the majority of the energy used in the UK economy is not used as electricity, but as fuels that power machines and industrial processes. Therefore any discussion about the merits of nuclear versus gas-fired generation is ignoring the greater issue – how will we replace these fuels?

Theoretically many of the uses of fossil fuels could be replaced by electricity. In the transport sector electric cars could be charged from the national grid, but as electricity is the second smallest form of direct energy consumption in the UK replacing the energy produced from petroleum and gas with electricity would require a massive expansion of electricity generating capacity. Also, as electricity generation is often less efficient than direct use of gas or petroleum, the primary energy consumption is likely to increase overall – converting from fossil fuels to electricity would not necessarily save much energy.

Even if we don't try and replace a large part of the energy sourced from petroleum or gas with electricity, the structure of the electricity industry itself needs a thorough review. It's been built-up over the space of sixty years in an environment where fuel supplies were plentiful and climate change was not an issue. Redesigning the industry to deal with these two restrictive factors will require a fundamental rethink of the device that forms the heart of our electrical power system – the *national grid*.

The development of the national grid began in the early 1930s. It allowed large power stations with high capacities to replace small local power generators. Following the Second World War the development of national grids allowed a large increase in the amounts of power generated and supplied. However, by relying on larger power sources the development of the national grid has made it more difficult for smaller power sources, like renewable energy, to compete economically with large-scale production.

It is a property of all materials that conduct electricity that they resist the flow of current to a varying degree. Resistance is a problem because it means some of the energy within the electric current will be lost as heat, and as this energy is lost, so the voltage in the wire will drop – according to *Ohms Law* (see Box 17). The implication of Ohms Law is that the voltage, and hence the energy lost in the wire is inversely proportional to the size of the current. And as the current is proportional to voltage, the higher the voltage, the less the drop in voltage that will take place over the same wire. So by increasing the initial voltage we can reduce the voltage drop, and consequently the energy loss. But this still leaves the problem of having 275,000 volts entering everyone's homes! For this reason the national grid uses a hierarchy of distribution lines, feeding power at different voltages to different users.

Box 17. Ohms Law and the Operation of the National Grid

The electrical resistance of a length of wire, and the voltage drop (V) this resistance causes, varies according to the level of the current (I) and the resistance of the material (R). The relationship between these three is known as Ohms Law. The voltage drop (in *Volts*) is equal to the current in the wire (in *Amperes* or Amps) multiplied by the resistance of the wire (in *Ohms*).

Let's say you want to send one kilowatt (1,000W) of power down a wire fifty kilometres long to a consumer, and the resistance of the wire is 0.5 Ohms per kilometre (R/km) To work out the current produced by 1,000W of electrical power we divide the power (Watts) by the initial voltage (V) in the wire. This produces a value for the current, which multiplied by the resistance of the wire gives us a figure for the voltage drop. Consider these three situations:

- The voltage at the beginning of the wire is 240V. This makes the current [1000W ÷ 240V =] 4.2A. The voltage drop will be [50km x 4.2A x 0.5R/km =] 105V, and the voltage at the end will be [240V − 105V =] 135V, and 43% of the energy from the power station would be lost.
- The voltage at the beginning of the wire is 11,000V. This makes the current [1000W ÷ 11,000V =] 0.091A. The voltage drop will be [50km x 0.091A x 0.5R/km =] 2.3V, and the voltage at the end of the wire will be [11,000V − 2.3V =] 10,997.7V, and 0.02% of the energy from the power station would be lost.
- The voltage at the beginning of the wire is 275,000 volts. This makes the current [1000W ÷ 275,000V =] 0.0036A. The voltage drop will be [50km x 0.0036A x 0.5R/km =] 0.091V, and the voltage at the end of the wire will be [275,000 − 0.1V =] 274,999.9V, and less than one-millionth of the energy from the power station would be lost.

As demonstrated by these three examples, using a higher voltage reduces the current, and as a consequence of the lower current the level of energy loss due to the resistance of the wire is reduced. This is why the national grid uses high voltages to move electrical energy around the country.

In order to divide the grid into different voltages it is necessary to convert the voltage level up and down. However, it wouldn't be safe for consumers to have 275,000 volts coming into their homes, and for this reason the grid uses *alternating current*, or AC (see Box 16), at a variety of voltages depending on how far the electrical power is being moved. With AC the voltage can be stepped up or down using transformers with only a small loss of energy. The result of this is that the consumer can be sure that the voltage they receive will be consistent − it will not vary depending how far away they are from the power station.

Each part of the electricity grid is separated from the next by switching stations and transformers (see figure 31). These step up or down the voltage for the next leg of the journey. This is how the national grid can deliver precisely the same voltage to consumers across the country. The sections of the national grid that cross the country – the *interconnectors* – operate at 400,000 volts. The interconnectors mainly take power from North Wales and Merseyside down to the West Midlands and London, and from East Anglia into London [POST, 2001]. From this main grid extends a network of high capacity lines – the *regional distributors* – running at 275,000 volts. These link to large power stations, and to regional grid connection points where the regional electricity supply companies source their power from the grid.

The main distribution system of the regional electricity supply companies runs at 132,000 volts. These link to major electricity users, such as heavy industries like steel producers or car manufacturers, and to area substations that provide power to urban areas. From these substations many lines radiate out operating at 11,000 volts. Some of these go directly to light industry and large commercial premises. Most run to local electricity substations that provide the standard mains voltage to homes and small businesses. Some small businesses that use slightly more power, such as on dedicated industrial

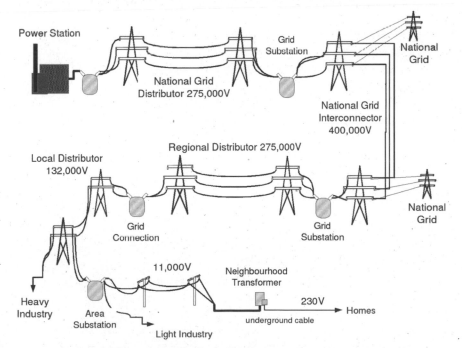

Figure 31. The National Grid and Regional Distribution System

estates, may receive power at an intermediate 415 volts.

The resistance of the wires in the national grid mean that some of the power is radiated as heat. The stepping up and down of the voltage also creates heat. The whole national grid and major regional distribution lines lose about 7% of the power they carry as heat or induced magnetic fields.

As well as the issue of how the grid works, there are also issues about how it has been managed. There is an argument that privatisation, and in particular the pricing arrangements for electrical generators, have led to a run down of plant and the capacity of the grid. It is argued that the price of electricity to the consumer doesn't reflect the economic price of production, and consequently the infrastructure of the industry is being slowly run down. If the continued pressure for low prices continues than it is possible that system failures could become more common [JESS, 2003], potentially leading to widespread power blackouts [BBC, 2004k] because the available generating stations couldn't meet demand [POST, 2003]. The problem is that because of the already poor state of the energy industry, and the pressure from the regulator not to pass on to customers the higher prices related to carbon trading and rising gas prices [ILEX, 2003], the industry may be unable to respond to the need to create more sustainable power generation and supply systems. So, even in the absence of Peak Oil and climate change, the UK's electricity generation system has its own home-grown crisis to deal with.

7. Low Carbon Energy Sources

The general definition of *renewable energy* is a source of energy that can continually flow from the natural environment. In contrast, conventional sources flow from the *stocks of fuel* that we hold, and when those stocks are gone the flow stops. The limiting factor to the use of renewable energy is that there are natural limits to how much energy is available to us over any period of time. Exceeding those limits is either impossible, because there's no more energy to harvest, or because doing so would destabilise other natural systems that rely upon that energy.

There are a number of truly renewable energy sources — these are dealt with in the next chapter. This chapter looks at the half-way house technology that is developing between conventional and truly renewable energy sources — *low carbon*. "Low carbon" because, compared to producing the same amount of energy from fossil fuels, these sources produce a lot less carbon dioxide, or the process only emits carbon sourced from the atmosphere rather than fossil carbon. This is an area where governments are placing a lot of support at the moment, mostly supporting the development of systems to process energy crops — crops that can be processed to produce energy as heat, liquid or gaseous fuels, or electrical power.

Despite the fact they are not strictly renewable, although many portray them to be so, energy crops offer an amazing benefit to the energy economists of the industrialised states — *people needn't change their habits*. Energy can be produced, and even liquid fuels can be produced, to substitute for the fossil fuels sources that we can no longer use. If this sounds too good to be true, it is. The flaw in the logic of energy crops is the *Law on the Conservation of Matter and Energy*. Energy crops do not grow energy. Instead the crop transforms the energy of the Sun, using the nutrients they extract from the soil, into *biomass* — biological fuels. Consequently the achievable level of energy production will always be limited by the amount of land that is available to grow the crop, and the amount of fertiliser that you have available to spread on the land to increase the crop yield.

Another area where governments and industrialists are investing a lot of effort is the development of systems that can use hydrogen as a fuel — the so called *hydrogen economy*. The problem with hydrogen is that it is not produced naturally as a renewable resource, and it's not found as a naturally occurring deposit like fossil fuels. Therefore hydrogen is not a fuel but, like electricity, a means of *carrying* energy. To create it you must first put energy into the system you use to produce the hydrogen. So as a potential resource hydrogen

Box 18. The Biomass Cycle – From Sun to Soil

Organic materials contain energy in the form of *carbohydrates* – compounds made of carbon, hydrogen and oxygen. The plant takes water (which contains oxygen and hydrogen) from the ground, carbon dioxide (containing carbon and oxygen) from the air, and solar energy (sunlight) to produce chemical energy using *photosynthesis*. This produces compounds called *monosaccharides* (simple carbohydrates), and an excess of oxygen which the plant returns to the atmosphere. By further stages of photosynthesis the plant polymerises these simple carbohydrates (in a process similar to the production of plastics from monomer compounds) to produce polysaccharides (complex carbohydrates), proteins, and by another level of processing, lipids (fatty compounds). The plant uses these different compounds to grow, creating *biomass*.

When biomass is burnt the opposite of photosynthesis takes place. At high temperatures the carbohydrates and other compounds produced by photosynthesis break down and react with oxygen in the air. This releases the chemical energy, stored in the biomass by photosynthesis, as heat.

Biofuels, like methanol or biodiesel, are created by physically processing the plant matter. Alcohol compounds are created by fermentation of the carbohydrates. Esters, the compounds that form the basis of biodiesel, are produced just from the fatty compounds of plants by extracting the oils in a press and then processing the oil to make it more fluid.

When plants are eaten by animals the carbohydrates, lipids and proteins are broken down by the process of *respiration*. This liberates energy as heat, and a range of nutrients that the animal uses to maintain and grow its own body. But as the digestive system of animals does not liberate the whole quantity of energy put there by photosynthesis there is still a lot of energy there that can be extracted from animal wastes.

To produce biogas the biomass must be processed using simple organisms. When biomass decays complex organisms like fungi, algae and moulds break down the plant matter to produce nutrients and energy. This produces an organic goo similar to animal excreta. To complete the cycling of energy and nutrients in the environment bacteria feed on the waste products to liberate the remaining energy. This produces a range of carbon compounds and other raw nutrients that plants use to grow new biomass.

There are two classes of bacteria.

Aerobic bacteria use oxygen to break down the material, creating carbon dioxide. The oxygen atoms are used to exchange energy in chemical reactions, and the cascade of chemical reactions initiated by the bacteria breaks down the carbohydrates (this is what happens in a compost heap).

Anaerobic bacteria work without oxygen, and utilise hydrogen atoms to transmit energy. This produces hydrogen-rich compounds. So instead of producing carbon dioxide anaerobic bacteria produce mainly methane, hydrogen sulphide, and some free hydrogen.

When organic matter rots in a confined space, like a landfill site, the immediate stages of decomposition are aerobic – producing carbon dioxide and water. That's because there's still a lot of free oxygen within the waste. As the oxygen is used up the anaerobic bacteria take over the decomposition process. To begin with *acidogenic* bacteria decompose carbohydrates by hydrolysis and fermentation, producing carbon dioxide, a little hydrogen and compounds like alcohols and acetates. The alcohols and acetates in turn form the input to *acetogenic* bacteria that break down the carbon compounds further, producing hydrogen. Finally *methanogenic* bacteria take hydrogen and carbon compounds and produce methane. It is the mixture of gases produced by anaerobic bacteria, mostly methane, carbon dioxide, and a little hydrogen and hydrogen sulphide, that we call *biogas*.

has the same problems as electricity generation – where do we find the energy required to create it, and how efficient are the conversion processes we use to make it.

There are other processes available that produce low carbon energy, but these are often reliant upon extremely unsustainable activities – like the production of waste. Let's begin with processes that produce *biogas*. This includes the digestion of organic wastes (see Box 18), such as sewage, to produce gas, and also the production of gas from landfill sites. These clearly fail the definition of renewable energy. This is because they rely on stocks of human-produced waste, such as household rubbish or catering and agricultural wastes, to produce energy.

Landfill sites waste resources, burying and contaminating them in a way that puts them beyond further use. However in some countries, and perhaps in this country in a few decades time, landfill sites are being mined to recover the valuable resources (mainly metals) they contain. Apart from the liquids leaking from the site, which can contaminate ground water and nearby watercourses, the landfill site produces gases as the material inside decomposes. This is mostly made up of the greenhouse gases methane and carbon dioxide, although the chemical wastes in the landfill will add a variety of toxic compounds to the gas stream. The production of methane is a particular problem because it is a far more potent greenhouse gas than carbon dioxide. So, energy production aside, converting the methane produced by landfill sites into carbon dioxide by burning it as an energy source is sensible because it reduces (minutely compared to the scale of the UK's emissions) the impact of waste disposal on climate change.

Modern landfill sites are sealed in an attempt to prevent their contents polluting the environment, and so the gases produced must be managed otherwise the seals would burst under gas pressure. The gas is collected from wells bored into the compacted waste. Then a network of pipes carries the gas to a flare tower, or on larger sites a landfill gas generator plant. Before it is used the gas must first be scrubbed because it contains impurities which would corrode the generator plant. The scrubbed gas is burnt in gas engines, or small gas turbines, that turn a generator to produce electricity. This produces mostly carbon dioxide, but also a complex mixture of pollutants because of the variety of chemical compounds that contaminate the landfill gas.

In 2003, landfill gas plants produced 46PJ (or 12.7TWh) of electricity [DTI, 2004g]. That's just over 2% of the UK's electricity production, and 31% of the UK's entire production of "renewable" energy for that year.

It's important to note that the energy produced from landfill gas, like incineration, does not make up for the loss of energy when recyclable materials are disposed of in the landfill. As a measure to reduce greenhouse gas emissions it is valuable, but it would be far better to eliminate the landfilling of mixed recyclable and organic wastes in the first place. Also, landfill gas is mainly produced by rotting organic matter – such as food waste, sewage sludge and paper. Other materials which might have been recycled, such as plastics and metals, do not produce landfill gas. As an energy option landfill gas is a very poor method of recovering energy from the waste stream. If only inert material were deposited in landfill sites, following on from intensive recycling processes, it would eliminate the large-scale production of methane, and would solve some of the ground and surface water contamination problems inherent with landfill.

A more energy-efficient alternative to landfilling is to conventionally recycle the plastics, metals and paper, but to take the organic fraction of the waste

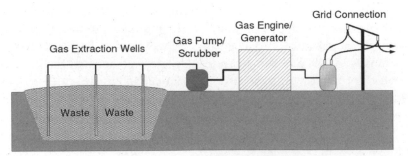

Figure 32. Diagram of a Typical Landfill Gas Power Plant

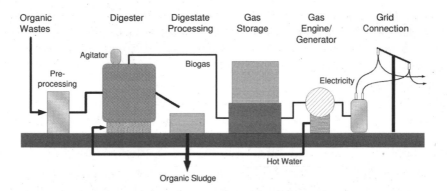

Figure 33. Diagram of a Typical Anaerobic Digester Plant

stream and break it down in an *anaerobic digester*. This is far more efficient than a landfill at breaking down organic matter to produce biogas. It is also a sealed process, which reduces the environmental impact and improves the efficiency of gas collection. Anaerobic decomposition processes are similar to those that take place in a landfill, but work at higher speed inside a sealed chamber heated to around 35°C (roughly body temperature). The processing of a batch of waste may take four to six weeks, and over this period biogas will be produced at varying rates. The overall volume of biogas produced is dependent upon the types of organic matter present in the waste.

Most anaerobic digester plants accept animal slurry or human sewage as well as organic wastes like commercial food processing waste. This is processed to produce the organic soup by maceration – blending the material with water to produce a suspension of very fine particles. The soup is activated by adding a bacteria culture to it, and then it is fed into the digester. Whilst in the digester the soup is stirred, or agitated, regularly. This ensures that the heavier particles of material are kept in suspension and can be digested by the bacteria. The bacteria digest the material, producing biogas composed mainly of methane, carbon dioxide and hydrogen sulphide. This is cleaned to remove the water vapour, stored, and then burnt in a combined heat and power plant – for example a gas engine – to produce power for the electricity grid and heat to run the digester plant.

After some weeks the processed organic matter can be removed from the digester and conditioned before disposal. This usually involves de-watering (settling and pressing) the slurry to make it more solid, and perhaps composting it for a short period to reduce the biochemical activity within the waste mass. In addition to the organic sludge, which can be ploughed into fields as a soil improver, the liquid produced by de-watering can be used directly as a liquid fertiliser [Biogen, 2002]. In the same way that soil bacteria return essential nutrients to the soil, so the material produced by anaerobic digestion contains the nitrates, phosphates and other essential nutrients that plants need to grow.

How efficient energy production from anaerobic digestion is depends upon how it is designed. Some processes, such as those processing human sewage, might add stages of heat treatment to kill-off pathogens (harmful bacteria or viruses) to make the organic sludge produced sterile. Others that work on more homogeneous waste streams may use little additional processing, and so produce a higher energy output [Biogen, 2002]. For example, Holsworthy Biogas in Devon processes 146,000 tonnes of animal and catering waste a year, producing 6,000,000m^3 of biogas [Farmatic, 2002]. The gas is burnt in a 2.1MW combined heat and power plant that produces heat for the digester and 14.4GWh of electrical power per year, of which 13.5GWh is sold to the national grid. This is a comparatively low level of output because the plant heat-treats the wastes coming into the plant to kill pathogens, which allows the digestate produced by the plant to be returned for agricultural use without the risk of communicating animal diseases. Scaling up, if the UK's 80,000,000 tonnes of agricultural wastes and 35,000,000 tonnes of sewage sludge [HMSO, 2000] were processed in the same type of plant then it would produce 10.6TWh/year of electricity, roughly 3% of the UK's electricity production in 2002.

The problem with anaerobic digestion is that, because it's essentially a waste technology, it's largely ignored by the Department of Trade and Industry. In the recent Renewable Innovations Review report [DTI, 2004f], produced to look at how the UK might meet the UK's 2010 target for greenhouse gas emissions, there was no detailed analysis of the potential of anaerobic digestion to produce power and displace carbon emissions. Likewise, the recent DTI report looking at the market for renewables, *Renewable Supply Chain Gap Analysis* [DTI, 2004g], also ignored anaerobic digestion. Even so, one of the consultants' reports commissioned by the DTI to produce the Renewable Innovations Review report did highlight the cost benefits and small-scale of operation of anaerobic digester plants, and noted that sewage sludge processing alone could provide 0.3% of the UK's electricity demand in 2020 [E4Tech, 2003]. This lack of interest is most likely due to squabbling between government departments, as the issues relating to sewage, waste and agricultural processes is the territory of the Department of the Environment, Food and Agriculture (DEFRA).

One energy source that the DTI's Renewable Innovations Review does give a lot of space to is biomass – utilising organic material directly to produce energy. Across the world biomass, in the form of firewood, is a major source of energy – supplying 11.3% of the world's energy resources. But the modern,

Box 19. How Much Energy Can Biomass Produce?

If we take the Eggborough plant as an example, the net energy value (energy output less the energy used for drying the wood chips) of the dried wood chips is roughly 5GJ/tonne. Short rotation coppice produces around 27.5te/ha (tonnes of biomass per hectare) [LEK, 2004], which at 5GJ/te (giga-Joules per tonne) equates to an energy output of 137.5GJ/ha (giga-Joules per hectare). But to grow the biomass takes three years, and so maintaining that power output takes three times that amount of land, reducing the power produced to 45.8GJ/ha/year – which at a 40% generation efficiency is equivalent to equates to 5.1MWh/ha/year of electricity†.

To produce just 1% of the UKs current electricity consumption would require 784,313 hectares of land to be planted with energy crops‡. That's equivalent to 18.5% of the 4,238,000 hectares [DEFRA, 2003] of land in cultivation in the UK, or 3.2% of the UK's entire land area.

The graph below shows what proportion of the UK's primary energy supply, or electricity production, would be produced if a certain percentage of the UK were covered in woody biomass.

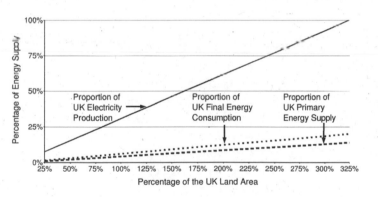

Figure 34. Energy Production from Woody Biomass

† 5GJ/te x 27.5te/ha = 137.5GJ/ha. Dividing by three years worth of growth equals 45.8GJ/ha/year. Then divide by 3.6 to convert from giga-Joules to mega-Watt-hours, and multiply by 0.4 (40% generation efficiency) equals 5.1MWh of electrical power per hectare per year.

‡ The UK's electricity production is 400TWh /year, or [400TWh x 1,000,000] 400,000,000MWh. 1% of this is [400,000,000 x 0.01] 4,000,000MWh. The area of land required to produce this is[4,000,000MWh/year ÷ 5.1MWh/ha/year =] 784,313ha (hectares). Therefore 1% of the electricity supply takes [784,313ha ÷ 24,290,000ha =] 3.2% of the UK's land area. Expressed as fraction of the UK's cultivated land it is [784,313ha ÷ 4,238,000ha =] 18.5%.

industrialised concept of biomass is the growing of fast-growing energy crops, such as willow, to produce the feedstock for gasification plants. In the same way as coal can be gasified in IGCC plants (as outlined in the previous chapter), organic matter can be gasified to produce a stream of flammable gas (mainly hydrogen and hydrocarbons) that can be burnt to produce power. This leaves behind a charcoal-like substance that could be burnt in a conventional boiler to produce heat or steam. At the moment plants that carry out gasification from biomass plants would need to produce ten megawatts, or more, of electrical power to make the process economic. In smaller plants, chipped forestry waste or bales of straw and other low-density plant matter can be burnt directly, as in other combustion processes, to produce heat and power.

To date there has only been one large-scale biomass plant developed in the UK – at Eggborough in North Yorkshire. This was a 10MW plant that intended to gasify 18,600 tonnes of forestry waste and 43,500 tonnes of biomass sourced within a 45km radius of the site. This would have produced around 80 to 85GWh of electrical power per year from a combined heat and power plant, the heat being used to dry the biomass before use in order to improve gasification. However, because of contractual problems and the failure of its backers to provide additional funding, in mid-2003 the developers of the plant went bankrupt just a few weeks after the plant went into full operation. Currently the plant is being broken up by an American company and shipped to India for reassembly and operation there [Guardian, 2003].

From the various reports available it is clear that biomass could be a useful source of energy. The energy the plants absorb from the Sun can be released directly through combustion or gasification. But there is one factor that these reports don't seriously consider – the amount of land it takes to produce significant quantities of biomass material in order to reduce dependence upon fossil fuels. Box 19 provides a calculation as to how much land the production of energy with biomass would take. These figures are highly dependent upon how much energy is contained in the biomass, and how efficient the energy transformation process is. Assuming we were to use a plant like Eggborough, to produce the same amount of energy (using CHP) as the UK's current final energy supply in 2003 (7.2EJ) would require 10 times the land area of the UK, but to produce the UK's primary energy supply (10.3EJ) would require 14 times the land area of the UK.

The fact that biomass produces a small amount of energy per hectare of land is an indication of the energy density of fossil fuels. These fuels represent the Sun's energy absorbed by an area of land (or sea) for thousands of years, compacted into a dense mineral by geological processes lasting millions of years. Consequently trying to produce energy from the Sun's energy input today produces only a fraction of the energy that fossil fuels yield. So whilst biomass is a useful source of energy, to make it a practical source of energy we would have to reduce our energy consumption significantly.

This point is true of another energy source produced from energy crops – *biodiesel*. Like biomass, biodiesel has obvious applications in a world that has little oil to fuel mobile energy uses, like cars or trains. As described in Box 18, photosynthesis provides the energy for the production of a whole range of chemical compounds. When plants polymerise simple carbohydrates to produce complex carbohydrates one of the compounds produced is *glycerol*. Glycerol forms the building block for a variety of fatty acids, and different types of plant produce varying quantities of these fatty acids. Rapeseed, or canola, is usually grown for the production of oils for the food industry. In hotter climates other crops are more suitable for oil production, such as maize (corn), olives or sunflowers.

The production of 1 tonne of biodiesel takes 2.8 tonnes of conventionally grown rapeseed [Hallam, 2003]. As a by-product this process creates 2.8 tonnes of rape straw (potentially useful as biomass), 1.6 tonnes of processed rape meal (which can be used in other industrial processes like soap making) and 100kg of crude glycerine. Biodiesel has a calorific value of around 37.3GJ/tonne (for fossil diesel it's 45.6GJ/tonne). The majority of the rapeseed grown in the UK contains a low quantity (less than 1% of the oil produced) of a fatty acid called erucic acid, which is indigestible to humans. A small amount of the rapeseed crop, roughly 50,000 to 60,000 hectares per year, is produced for industrial uses and contain 50% to 60% erucic acid by volume of the oil produced. Oil high in erucic acid can be processed by refining and esterification to produce rapeseed methyl ester – the compound that forms biodiesel. Box 20 gives calculations for how much land is required to produce biodiesel.

The oils produced by energy crops, like rapeseed oil or sunflower oil, can be used directly in a diesel engine. In fact, when the diesel engine was first invented it was run on vegetable oils (only later was diesel produced from fossil oil). This requires that the diesel engine is converted – replacing the fuel injectors and adding a system to heat the oil – because vegetable oils are more viscose than ordinary diesel. For this reason processed biodiesel is an easier option because it doesn't require conversion, although the extra processing and transport means that it is more expensive. The compromise between the two is a "two tank" system. The vehicle is started using fossil diesel or processed biodiesel, but then raw oil is added to the fuel mix when the engine has heated up. It is also possible to run diesel engines entirely on vegetable oil with a conversion kit [Tickell, 1998].

The discussion of energy crops should hopefully allow you to understand the two factors that are the major flaw in any argument that proposes that energy crops could replace the use of fossil fuels:

Box 20. How Much Energy Can Biodiesel Produce?

Intensively produced rapeseed yields 3.1te/ha (tonnes per hectare), or 41.4GJ/ha of energy[†]. If we take a diesel car that travels 45 miles per gallon, that's an energy expenditure of 0.0039GJ per mile. So one hectare of land will fuel a diesel car for 10,615 miles. Or, if the average car travels 9,000 miles per year, each car would use 0.85 hectares of land per year (or roughly two acres per year)[‡].

Scaling up, the 50,000 hectares of rapeseed high in erucic acid that are grown in the UK would support just 58,972 diesel cars for one year. The UK's entire rapeseed crop, 357,000 hectares in 2002 [DEFRA, 2003], if grown as rapeseed high in erucic acid would keep just 421,060 cars on the road for a year[§]. This assumes that the same high intensity cultivation can continue in the future. If we reduce the fertiliser inputs then the yield would drop to 2.9 te/ha, and the quality of the oil drops so it takes 3 te of rapeseed to produce 1 te of biodiesel [Hallam, 2003]. In this case the average car would require 1.0ha of land per year instead of 0.85ha/year.

Figure 35 shows how much of the UK's land area would need to be covered with rapeseed crop in order to keep a certain number of vehicles on the road, depending upon the level of fertiliser input.

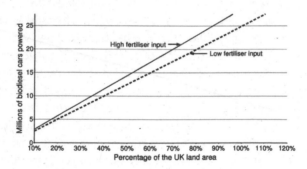

Figure 35. Energy Production from Biodiesel

[†] Producing 1te of biodiesel takes 2.8te of rapeseed. 1ha will produce [3.1te/ha ÷ 2.8te =] 1.11te/ha of biodiesel, therefore the energy density is [1.11te/ha x 37.3GJ/te=] 41.4GJ/ha/year.

[‡] Fossil diesel has a density of 260 gallons/te and calorific value of 45.6GJ/te. The energy used is [45.6GJ/te ÷ 260 gallons/te ÷ 45mpg =] 0.00390GJ/mile. 1ha of rapeseed will power a car [41.4GJ/ha ÷ 0.0039GJ/mile =] 10,615 miles. At an average 9,000 miles/year, each car requires [9,000 miles/year ÷ 10,615 miles/ha] 0.85 ha/year.

[§] 50,000ha/year of oilseed multiplied by 10,615 miles/ha equals 530,750,000 miles/year. Divide by 9,000 miles/car/year equals 58.972 cars. For 357,000ha of oilseeds the result is 421,061 cars.

Firstly, energy crops do not grow energy – that is impossible, as specified by the *Law on the Conservation of Matter and Energy*. Instead the crops transform the energy supplied by sunlight into biomass, utilising the nutrients they extract from the soil in order to undertake the necessary chemical processes involved. Therefore the production of energy crops will always be limited by the availability of land, and by the availability of nutrients in the soil to enhance the productive capacity of the land.

Secondly, fossil fuels are a very dense form of energy. They contain a very high level of energy per unit of volume, but that energy is actually a fraction of the sum of the energy imparted to a large area of land by the Sun over thousands of years. Therefore energy crops can never practically replace fossil fuels because they cannot produce the scale of energy production fossil fuels represent without the use of huge areas of land. So at a time when climate change is likely to be reducing the amounts of high quality agricultural land across the globe, expecting that we can source energy from energy crops is unrealistic unless we are prepared to seriously inhibit the production of food.

Biomass and biodiesel are likely to be an important source of energy for uses where other energy sources would be inefficient or difficult to utilise. For example, biodiesel would be an option for transport in remote rural areas where the use of mass-transit systems would be inefficient in energy terms. Likewise, biomass could make a useful back-up to other unpredictable sources of energy such as wind and wave power, particularly if used as a source of combined heat and power (which increases the overall efficiency). However, neither will be able to supply the amounts of energy that society has been using annually over recent history. All routes to the production of fuel oils and gases, like combustion, gasification, the production of biodiesel, and the *pyrolysis* (heating to high temperatures in a sealed vessel without oxygen) of biomass to produce gas and oils, is still limited by the *Law on the Conservation of Matter and Energy* and the *Second Law of Thermodynamics* – you can't get more energy out than the plant was able to store from the Sun, and in reality you will be lucky to get six-tenths of the energy content of the biomass.

Like energy crops, hydrogen is attractive to the energy establishment because it replicates many of the features of existing energy systems. It is a potent source of energy, and it can be processed to produce electricity and/or heat – making it ideal for high efficiency combined heat and power plants. The only problem with hydrogen is that, although it may be the most plentiful element in the universe, we don't have lots of it lying around as free hydrogen gas. Hydrogen must be manufactured, and for this reason it can only ever work as an energy carrier, like electricity. This of course runs us once again into the *Law on the Conservation of Matter and Energy*. We can't create hydrogen from nothing, and when we do manufacture hydrogen the production process will yield less useful energy than the energy invested in it. In the same way that the low efficiencies implicit in the generation of electrical power affect the level of production, the efficiency of hydrogen as an energy system is dominated by the efficiency of hydrogen production and supply.

Despite these restrictions on the use of hydrogen, there is a lot of noise coming from some governments, and especially the car industry [Guardian, 2004q], about the importance of hydrogen as 'the fuel of the future'. For example recent research from the UK [Oswald, 2004] suggests that to convert the UK's car fleet to hydrogen would require the construction of 100,000 wind turbines, or 100 nuclear power plants [Warwick, 2004].

Hydrogen bonds with oxygen energetically. Consequently when you react two atoms of hydrogen with one atom of oxygen you get water and a release of heat and electrons. This has been known about for some time. Sir Humphrey Davy demonstrated the electrolysis of water in 1807. Then in 1850, Sir William Grove invented the *fuel cell*, a device that could efficiently reverse this process by recombining hydrogen and oxygen to produce water and an electric current. In fact, the principle was so well known in Victorian Europe that in 1874 Jules Verne, in his novel *The Mysterious Island* [Verne, 1874], described the day when, "water will some day be employed as a fuel".

Hydrogen is produced by four common processes:

- *Reformation of hydrocarbons with steam* – heating hydrocarbons such as oil or gas to high temperatures (around 900°C) and then injecting high pressure steam to react the hydrocarbons and produce carbon monoxide and free hydrogen.
- *Gasification or partial oxidation* – heating coal or oil, and then injecting steam, to liberate the hydrogen – this is roughly the same process that forms the basis of Integrated Gasification Combined Cycle (IGCC) power stations, but instead of burning the gas in the turbine it is processed to extract the hydrogen.
- *Pyrolysis* – hydrocarbons and biomass can be heated to high temperatures, in the absence of oxygen, so that the compounds break down to produce carbon compounds and free hydrogen.
- *Electrolysis* – an electric current is passed through water to break it down into hydrogen and oxygen.

Of these the steam reformation and gasification of hydrocarbons are likely to be the cheapest in the short term. Of course, these options require that we have cheap hydrocarbons available to fuel the process. These processes also emit a lot of fossil-based carbon dioxide which would have an equally significant impact on climate change as burning the fossil fuels directly. Eventually either the scarcity of hydrocarbons, or the need to control the emission of carbon dioxide, will make these processes more expensive than the alternatives.

Pyrolysis is the dirtiest process because the heat and pressure that frees the hydrogen also enables countless other chemical reactions to take place which create a wide range of hydrocarbon and aromatic hydrocarbon compounds. Electrolysis is the cleanest, and has minimal emissions, but it is the most expensive because you must provide all the energy required for hydrogen production.

When the hydrogen has been produced a second problem arises – storage and distribution. Hydrogen creates a lot of energy when it reacts with oxygen, which is why it explodes so vigorously. Although per unit of mass hydrogen supplies more energy than petrol (120MJ/kg for hydrogen over 42MJ/kg for petrol), in terms of the volume of gas hydrogen delivers little usable energy. To make it compact enough to use hydrogen must be compressed, and this requires energy to accomplish. The fact that it has to be stored at high pressure also means that hydrogen cylinders have to be strong, which requires a lot of heavy metal. This affects the efficiency of distribution because as well as transporting the hydrogen fuel you must also transport the weight of the container it is held within. Even when it is compressed to 200 times atmospheric pressure (a usable storage pressure for hydrogen) 1 litre of natural gas still has more energy than 1 litre of hydrogen [POST, 2002]. But, it is argued, the fact that it can be utilised at a 40% or 50% level of efficiency means that the losses from its low energy yield are made up by the higher efficiency when compared to the efficiency of petrol or electric cars. Another option for hydrogen storage is to absorb the gas onto metal alloys to produce metal hydride – for example using alloys of nickel. But this option is still expensive, and still requires a lot of weight to be moved around.

If you burn hydrogen to produce heat its efficiency of use is no better than other flammable gases. The most efficient way of extracting energy from hydrogen is by using a *fuel cell* (see Box 21). As noted above, the principle of the fuel cell was discovered in 1850, but it was not widely used until the 1960s and 1970s when space vehicles needed a clean and compact source of electrical energy. In addition to electrical power, fuel cells also produce a certain amount of heat which can be removed and utilised with a conventional liquid cooling system.

An option to avoid the problems inherent in the storage and transport of hydrogen is to use a hydrogen rich compound as a *precursor fuel*. The fuel, which could be a liquid or a gas, is reformed to produce a waste compound and free hydrogen using some of the energy created by the fuel cell. The problem with this is that reforming the precursor fuel lowers the overall efficiency of the fuel cell. The most popular candidate as a precursor fuel, for mobile

Box 21. How Does a Hydrogen Fuel Cell Work?

Fuel cells produce electrical power by the exchange of electrons between molecules of oxygen and hydrogen as they react in the presence of the catalyst (see figure 36). When two molecules of hydrogen gas ($2H_2$) break down at the anode they produce four positively charged hydrogen ions ($4H^+$) and four spare electrons. When an oxygen molecule (O_2) breaks down at the cathode it produces two negatively charged oxygen ions ($2O^{-2}$) and a deficit of four electrons. This difference in charge between the anode and the cathode creates an electron flow, and consequently an electric current between the two terminals of the fuel cell. This leaves a collection of positive hydrogen and negative oxygen ions within the cell that migrate through the electrolyte to the opposite terminal. The role of the electrolyte is to allow the ions to move across the fuel cell without reacting.

When two hydrogen ions ($2H^+$) meet an oxygen ion (O^{-2}) in the region of the anode or cathode they react, with the help of the catalyst, to form one molecule of water ($2H^+ + O^{-2} = H_2O$). These reactions also produce heat which can be extracted, along with the water, from both terminals of the fuel cell.

Another possibility is that certain types of fuel cell, such as the reversible proton exchange membrane fuel cell, can operate in a forward direction (as a conventional fuel cell) to produce power, and in a reverse direction (working as an electrolysis cell) to produce pure oxygen and hydrogen from water. The benefit of this process is that, rather like a battery, you could charge the fuel cell plant by putting power into it and storing the gases produced – provided you have a system to collect the hydrogen and store it safely (such as a nickel-metal hydride absorber). Then when you need that energy you reverse the process, through the same piece of equipment, to produce power.

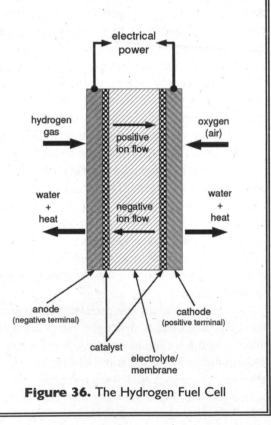

Figure 36. The Hydrogen Fuel Cell

applications at least, is a mixture of water and *methanol* (or *methyl alcohol*). This reacts to produce free hydrogen and carbon dioxide. The hydrogen is then absorbed into the fuel cell where it reacts as usual.

How efficient fuel cells are depends upon what type of fuel cell you use, and how you use it. Fuel cells that reform a precursor fuel usually have a low efficiency, of between 30% to 50%, because of the extra energy used in reformation. Those that use pure hydrogen may have an efficiency of 50% to 70%. This is why the method of fuel production is important. Whilst fuel cells may offer higher operating efficiencies, the use of energy to manufacture the hydrogen or precursor fuels, coupled with the efficiency of the fuel cell, may not create a vast improvement in energy efficiency over conventional energy sources.

Although there are clear problems with using pure hydrogen fuel cells, using a precursor fuel, such as liquid methanol or methane gas, would make the fuel cell far easier to use. The advantage of using a precursor fuel, like methanol, is that it can be carried as a conventional liquid that does not have the flammability or explosive properties of hydrogen. The downside is that you must make the methanol, usually using a fermentation process, and this can require a lot of energy. However, methanol can be produced by fermentation processes that use sugar-rich biomass as their feedstock. This avoids the problem of having to use energy sources created from fossil carbon, but you still have to find the lands to grow the sugar-rich biomass.

In static applications, fuel cells fed with natural gas make a far more efficient combined heat and power system than traditional gas-fired CHP. These are already being used as high-efficiency combined heat and power plants. For example Woking Leisure Centre in Surrey, where a fuel cell creates heat and power for a swimming pool, reforming natural gas from the gas main [BBC, 2003d]. It would take little effort to get similar systems operating with biogas plants, however the biogas would require special processing to remove any contaminants that might damage the catalysts within the fuel cell.

Obviously the fuel cell is a useful device. Whilst it may not in practical terms meet the hype many people attach to it, it could provide a valuable tool to manage and produce energy. However, there is one issue that could still cast a doubt over the widespread use of hydrogen as a fuel. In the the lower part of the atmosphere, the troposphere, free hydrogen is rare. It is so reactive that it is quickly scavenged from the environment and bound up as water vapour and other compounds. But at higher altitudes, in the stratosphere, hydrogen is equally scarce. This is because one of the major sources of hydrogen, water vapour, is excluded from the upper atmosphere by the tropopause – an area of intense cold at around 16km above the Earth's surface where temperatures reach −50°C. This freezes out the water vapour as ice crystals (forming clouds) and prevents it rising further than 8km to 10km.

If a large-scale hydrogen economy emerges, based on the use of hydrogen gas rather than precursor fuels, then leakage of hydrogen is inevitable. This could be as much as 10% by volume [Caltech, 2003a]. As hydrogen is so light, any leak would rise speedily through the atmosphere, and as it is free hydrogen the tropopause would not arrest its ascent. Although some of the hydrogen will be scavenged as it rises within the troposphere, a large amount will make it to the stratosphere. Here it will react with oxygen to produce water vapour. This will produce clouds at very high altitudes, affecting the albedo of the Earth and cooling the Earth slightly. Given the problem of climate change this might be welcome. But a large source of the reactive oxygen the hydrogen is likely to react with makes up the ozone layer, 30 kilometres up in the stratosphere. In the same way that chlorofluorocarbons (CFCs) scavenged ozone to create a hole in the ozone layer, so too will free hydrogen. This would lead to the same kind of thinning of the ozone layer created by CFCs, and the same increase in the flux of ultraviolet light that reaches ground levels to create health problems in humans and in natural ecosystems. What's more, the cooling effect create by stratospheric clouds would lengthen the seasonal cold period over the north and south poles during which the ozone holes grow. This would contribute to a deepening of the cycle that creates the ozone holes, making the overall problem worse [Caltech, 2003b].

So at one level the hydrogen economy may offer some benefits from the increased efficiency of combined heat and power plants. It may also provide a means for cyclical or unpredictable renewable sources to produce a constant power output. If, that is, we can find a means of efficiently producing hydrogen, or precursor fuels, from biomass. However, unless this is done in a way that minimises the leakage of hydrogen gas the potential benefits could be lost due to the damage to the ozone layer.

Analysing the benefits and dis-benefits of biomass, biofuels and hydrogen is complex. But it is clear that many of the problems with the use of all these fuels, both in terms of energy economics, and their practical impacts, are being ignored in the rush to implement low carbon alternatives to fossil fuels. A truly sustainable energy policy would seek to provide the most efficient use of natural resources, within the bounds set by the ecological systems of the Earth's atmosphere and ecosystems. Such a policy, by default, should also reduce greenhouse gas emissions. Energy crops and hydrogen, as a sort of half-way measure between conventional and renewable energy technologies, represent a fine line between what is an ecologically sustainable energy policy and what is a slightly more sustainable industrial energy policy. Whilst there is nothing intrinsically wrong with their use, which side of this line they sit upon depends on how these technologies are used and managed.

8. Renewable Energy

As noted at the beginning of the previous chapter, the general definition of renewable energy is a source of energy that can continually flow from the natural environment. However, the UK government takes a very wide definition, the result of a policy shift in the 1990s that tried to integrate waste disposal into energy policy [DOE, 1992]. So under the UK's statistics for renewable energy you don't just find wind, solar, wave or water power, but also waste incineration and landfill gas. In this chapter the strict definition is applied – the sourcing of flows of energy solely from the natural environment.

The media usually portray renewable energy, as a result of current national policy, as just generating electricity. But when we look at the whole range of potential renewable energy sources one other significant form of energy is available – *heat*. Either sourced directly from the Sun using thermal solar systems, or by extracting it from the environment using a heat pump. This seemingly ignored energy source is important because heat forms the majority of the energy used by those in temperate climates. In the UK, homes use half of their energy for space heating and one quarter for water heating. Consequently producing heat saves a larger quantity of energy than is actually used, due to the avoidance of the system losses when electricity is generated and converted to heat at the point of use.

Currently most of the energy the UK government classes as "renewable" actually bears little relation to the public's view of what constitutes renewable energy (see figure 37). Whilst we may have visions of wind turbines and solar panels, in 2003 most renewable energy came from landfill gas (31%), burning waste and car tyres (23%), and burning animal wastes (17%) [DTI, 2004e]. Of all the truly renewable sources, hydro was the largest at nearly 8% of renewable energy production. Wind power only produced 3% of the UK's renewable energy, and geothermal and active solar systems less than 0.6%.

There are four factors affecting the adoption of more renewable energy sources in the UK.

Firstly, *price*. The effect of deregulation in the energy sector has been to keep electricity prices low as a consequence of regulatory pressure. This, some argue, is actually damaging the supply infrastructure. It also restricts the ability of renewable energy to compete on economic terms with conventional power generators. For those wanting to install their own energy systems for electricity or heat, or to reduce their energy consumption through improving the efficiency of their appliances, or to reduce the heat loss from their home, low energy prices create a cost penalty. It lengthens the period

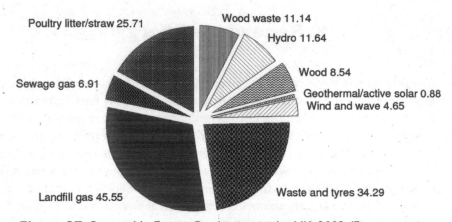

Figure 37. Renewable Energy Production in the UK, 2003 (Figures are in petaJoules) [Source: DTI, 2004e]

over which their improvement will pay back in energy savings, making change less worthwhile.

Secondly, *government policy favours the large-scale development of supply-side renewable energy projects, not the demand-side management of energy use.* Consequently, by focusing on substituting primary energy production, the potential savings from the more efficient use of energy are missed.

Thirdly, *the development of renewable energy systems in the UK is centred around the development of large grid-connected projects, usually carried out by large organisations.* The collective effect of promoting thousands of very small projects does not appear to be a priority within the Department of Trade and Industry because their focus is on corporate answers to energy need, not individual solutions for the needs of small energy users.

Finally, *there are few good examples of sustainable energy use promoted through the media* – all we see is green energy or environmentally friendly production. This affects the public's perception of what renewable energy is. The public see wind turbines presented as an icon or totem for renewable energy, when in fact, in terms of the public's demand for energy overall, a thermal solar panel would be more relevant to their everyday lives. But whilst we see these images in the media, the reality of the renewable energy in the UK means we should instead see images of landfill sites or waste incinerators.

Renewable systems extract energy from the environment. This means that the type of energy we collect has to be related to the types of energy flows that exist in the environment. The types of energy we can easily harvest are heat energy and kinetic energy (the energy of motion held within the wind or moving water). For this reason how we reap energy from the natural environment is restricted to the efficiency with which we can extract either heat or kinetic energy, moving or transforming it so that we can direct it to where it's needed.

The principle restriction on renewable energy systems, because they source energy from natural systems, is that they can only produce relatively small amounts of power. They do not have the energy density of fossil fuel sources. A few technologies, such as wind power, will scale-up. Most renewable energy technologies are far better suited to providing power for small energy users at the point of use, such as those sources that can efficiently produce heat (for example thermal solar or heat pumps) rather than providing large quantities of energy for supplying centralised distribution grids [OU, 2004].

The other significant restriction, which perhaps we are meeting for the first time with the backlash against industrialised wind power, is that there are natural limits to the use of renewable energy. This is either because of the technical restrictions of harvesting it, the practicality of the area of apparatus you need, or because taking too much energy would cause other natural systems to break down or be degraded. Renewable energy systems have as much potential to damage the natural environment as conventional energy systems if they are developed inappropriately or they are designed to take so much energy from natural systems that the local ecosystems are affected.

Although there are many ways we can get renewable energy, the variability in the natural environment makes some options better than others. Perhaps top of this list in terms of variability is *geothermal energy* – heat that is extracted from underground rocks. *Low-grade heat* percolates up from the ground and this can be intercepted by heat pumps (covered later in this chapter), but where volcanic activity or hot springs heat the ground a lot more heat, and even electrical power, can be produced. However, there are only a few regions in the world where using the high grade heat from geothermal sources is an economic option. One of the principle restrictions is that unless there is a substantial urban area over the hot rocks to use the hot water produced by these systems then the only option for using geothermal energy is power generation. To do this the temperature in the rock needs to be in excess of 150°C to 180°C, and there are only a few places in the world where the rocks are hot enough to do this. Within Europe, the only significant use of geothermal power is in Iceland and Italy.

There have been some experiments with geothermal energy in the UK, such as in Southampton where geothermal heat is used to heat water with a heat pump for a district heating scheme. The current restriction on use of geothermal technology is that it mainly works within sedimentary rocks, not the hotter volcanic rocks that hold the greatest amount of heat. This is because geothermal systems need a flow of water beneath the ground so that hot water can be pumped out, have its heat extracted and be pumped back down to the hot rock strata to be heated again. As volcanic rocks are not readily water permeable it is very difficult to get the heat they contain out. There is an ongoing European research project, the *Hot Fractured Rock* (HFR) *Pilot Project* [EHDRP, 2003], to try and develop heat extraction from dry rocks, and if perfected there would be potential for using it in the UK – mainly in northern England, Scotland, Northern Ireland, and in particular Devon and Cornwall (which have some of the hottest rocks in the UK).

Thermal solar is the name given to a range of systems that trap the energy radiated by the Sun. There are two forms of the technology:

- *Passive systems* receive energy from the Sun without the addition of any additional energy inputs that enhance the efficiency of collection.
- *Active systems* use technological measures, such as water pumps, to enhance the collection and transport of heat energy.

How much solar heat these systems can produce depends upon their design and location. A common misunderstanding about thermal solar energy is that the Sun must be shining. This is not the case. What carries the heat energy of the Sun is short-wavelength infra-red radiation. Although this can be attenuated by cloud, it is not stopped. So even on a cloudy day, unless the cloud is very thick and dark, solar energy can still be collected. Also direct sunlight isn't wholly required. The atmosphere scatters the infra-red light from the Sun so that, providing there's a clear view of the general area of the Sun, the collector will pick up a lot of heat radiation even when the Sun is not directly overhead. To help with the design of solar systems some organisations have produced maps of solar insolation [Solarex, 1996]. These give a figure for the minimum energy provided by the Sun, in kilo-Watt-hours of energy per square metre of solar collector per day ($kWh/m^2/day$) to a solar collector that is optimally positioned to receive energy. However the energy that you can actually extract from the sunlight will be less than this value, depending upon the type of collector you are using and the efficiency of the energy transfer and storage system.

The commonest form of solar power used in the UK is passive solar heating. The use of windows to increase the amount of sunlight that enters a house. This heats the house during the day by capturing heat inside the rooms, so reducing the need for other heat sources to be provided. This can be enhanced if a long wave infra-red filter is applied to the inside of the glass (commonly called *K-Glass*) because it allows the short-wave infra-red radiation from the sun to enter the room, but the re-radiated long wave infra-red radiation from the hot objects in the room can't escape. Another common passive solar device is the conservatory. Controlling the movement of the warm air from the conservatory allows the collected heat to be moved through the whole house. The large area of glass means the solar input is higher, and so it acts as a source of warm air during the Spring and Autumn, but can be closed off as required to prevent excessive cooling or overheating during the Winter and Summer.

During the 60s and 70s the 'solar house' was a popular design style, but it didn't truly harness solar energy. It had large areas of south-facing widows, and consequently they could be cold in Winter due to the increased heat loss from the large area of glass, and hot in Summer as the windows allowed a large input of heat to the rooms. True solar houses control the collection and absorption of heat very carefully, using solar heat in two ways. Firstly, like a conservatory, they use heat to warm air and circulate it throughout the building. Secondly, they also trap some of that heat energy in the fabric of the walls and floors. For example, the *Trombe wall*, which collects heat during the day and passively radiates it into the house at night. This means that as the level of energy input from the Sun falls during the evening this trapped heat can help maintain the internal temperature of the house. The overall effect of passive solar heat is to reduce the amount of fuel-sourced heat required to heat the spaces in the house, particularly during the Autumn and Spring when air temperatures are low but the days are long enough to provide prolonged direct sunlight.

The best collectors of solar heat are active solar systems. This is because a solar collector can be designed to absorb a large amount of heat energy and transfer it to a coolant, such as water. Infra-red radiation is preferentially absorbed by dark-coloured objects. The best colour is matt black, but specialised coatings are now available that specifically absorb short-wave infra-red radiation. Some of that energy will be reflected away by the surface. As the collector heats up it will also lose heat by radiation (as it shines long wave infra-red radiation) and by convection (as the air conducts heat away from the collector). The best solar collectors seek to avoid these losses in order to maximise the amounts of heat collected.

Flat plate solar collectors use a sheet of metal with a network of coolant pipes. This is enclosed within an insulated case with a glazed front to allow sunlight inside. More recent flat plate collectors have evolved various means of concentrating the light entering the box – such as the use of dimpled or ridged glazing that concentrates the light from a larger area rather than just a narrow band perpendicular to the collector. The most efficient thermal collectors are *evacuated tube collectors*. This is a narrow tube of glass with a thin metal plate running down the centre. The air inside the tube is removed, reducing thermal losses from the collector plate by convection, and the

Box 22. Heat Production from a Thermal Solar System

All substances have a *specific heat capacity*. This is a measure of how many Joules of energy it takes to heat 1 kilo of the substance by 1°C. For water, the specific heat value is 4,190 Joules per degree Celsius per kilo (J/°C/kg). Water enters buildings at around 8°C to 11°C. Heating 1 litre (1 kilo) of water to 60°C with electricity takes about 217,880J. Note, with gas heating operating at 50% to 60% efficiency, the actual energy expended to heat that water would be just under twice this amount (although overall gas is better than electricity because of the efficiency and transmissions losses associated with electricity generation).

Over a year a solar system might heat water to 60°C in Summer, but perhaps not at all in Winter. So on average the water will be preheated to 30°C. Heating 1 litre of water to 60°C from 30°C only takes about 125,700J[†] – a saving of 40% on average.

Sizing a solar collector for a particular application is dependent upon the amount of heat required, and the rating of the collector. A system designed to heat only a proportion of the hot water during the hottest Summer months would produce little energy for the rest of the year. Likewise a system designed to produce a lot of heat in Winter would provide far too much in Summer. So ideally a system is designed to produce about half the heating load required during the Spring and Autumn – it will then produce just about all the load in Summer. Ignoring the efficiency of collection and thermal losses let's say we want to preheat 300 litres of cold water over one day from 8°C to 35°C. In total that's 33.9MJ/day[‡].

If we take the daily insolation for Spring and Autumn as 2kWh/m²/day then producing 33.9MJ/day would require a solar collector with an area of 4.7m². Multiplying this average heat load over a year, and comparing it to the cost of electricity, this heat energy is worth £275 per year[§]. Even so, the current high cost of retrofitting solar water heating to older homes (in excess of £2,500) means that the pay-back period for these systems can be around ten to fifteen years. However, if these systems were installed as standard on new homes the scale of operation would make the extra costs negligible compared to the overall costs of new homes.

[†] Assuming an average distribution between the Summer and Winter heating load: 4,190J/°C/kg × (60°C – 8°C) × 1kg (or 1 litre) = 217,880J. 4,190J/°C/kg × (60°C – 30°C) × 1kg = 125,700J. (1 – (125,700J ÷ 217,880J)) = 42%.

[‡] The required daily input would be 4,190J/°C/kg × (35°C – 8°C) × 300kg (300 litres) = 33,939,000J, divided by 1,000,000J to convert to megajoules = 33.9MJ. 2kWh/m²/day is equivalent to 2kWh × 3,600 seconds = 7,200kJ/m²/day, divided by 1,000 to covert to megajoules = 7.2MJ/m²/day. 33.9MJ/day ÷ 7.2MJ/m²/day = 4.7m².

[§] As the design represents the average heat load, the total heat produced over a year would be 365 days/year × 33.9MJ/day = 12,373MJ. If we divide this by 3.6MJ to convert to kilo-Watt-hours, and then multiply by the cost of a kilo-Watt-hour of electricity (8 pence) the value of this heat would be (12,373 ÷ 3.6MJ) × £0.08/kWh = £274.95 per year.

profile of the plate inside the tube gives a wide angle for light collection. Evacuated tubes produce more heat, but they are more expensive, and there can be problems with their long term reliability because if they are broken or if the seals break at the ends of the tube (the internal vacuum will be lost and the collector doesn't trap as much heat).

The key figure in designing a thermal solar system is the *solar fraction* – how much of the annual hot water, and possibly , energy load will be provided by the solar system. As a trade off between efficiency and cost, many systems aim for a solar fraction of about 0.4 to 0.5. That is, 40% to 50% of the hot water load is produced by the solar panels. Less than this isn't really cost effective. More than this creates problems because in Summer the system will produce. more heat than the house can use – unless you build a *heat store*. Heat stores, usually a large tank of water, can be charged up with heat during the day using the excess of solar heat. This is then used for heating after dark, or when the solar input falls during cloudy weather.

The heat collected by thermal solar systems usually preheats the hot water supply before it is heated to the final temperature for use. It could be used to heat the air in the house but that would take a much larger collector, and a heat store would have to be built into the house where the energy collected during the day could be stored.

The solar heating system is a closed loop (see figure 38). Rather than heating water directly, it heats the water in the collector system and this passes through a preheat tank that functions as a heat exchanger. To circulate the water a pump is used. When the control electronics sense that the temperature of the solar collector is a few degrees above the temperature of the preheat tank the pump switches on, cooling the collector and heating the water in the tank. If the collector cools too much then pump switches off to prevent cooling the tank.

Another useful thermal technology are *heat pumps*. Just about every home

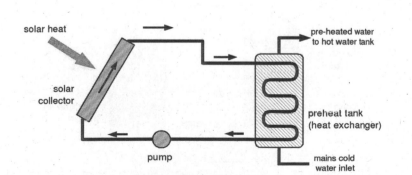

Figure 38. Diagram of a Typical Solar Water Heating System

has a heat pump in it, inside the refrigerator. This cools the inside of the fridge and dumps the excess heat out of a cooling plate at the back. Using heat pumps we can utilise renewable low grade heat from the environment by raising it to high grade heat (in short, making it hotter). You will remember from earlier in the book that the *Second Law of Thermodynamics* states that energy can only flow from a high state to a low state. However, heat pumps don't offend the Second Law because although they reverse the flow of heat, the penalty for this is that you have to put more energy into the process to get the value of the higher-grade energy that is produced. Even so, the heat saved by using heat pumps off-sets the use of heat energy, and hence saves energy overall.

A heat pump has four main components that together form a closed loop. The evaporator boils a liquid with a low boiling point, called the refrigerant, to form a gas. This creates a cooling load, removing heat from the low grade heat source it is placed within. Next the compressor sucks the gas from the evap-orator and places it under high pressure. As the gas flows through the con-denser under pressure it condenses to form a liquid and dumps its latent heat. This creates a heating load, producing high grade heat. Finally the liquid passes through an expansion valve or throttle so it can enter the evaporator at a lower pressure, evaporate to form a gas, and go through the cycle once again.

Ideally you are looking to extract as much energy from the evaporator by providing as little energy or work to the compressor as possible. A well designed heat pump system could cut heat demand by between 50% and 70%. To put in in monetary terms, if we put £100 worth of electricity into a (COP = 2.5) heat pump to extract low grade heat from the environment and put it into an under floor heating system, we would receive a free energy contribu-tion from our low-grade heat source equivalent to £250 worth of electrical power (see Box 23).

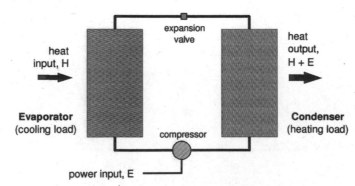

Figure 39. A Heat Pump System

Box 23. How a Heat Pump Works

Heat pumps use two properties of matter – their *boiling point*, and their *latent heat of vaporisation*. The boiling point of a liquid varies with pressure. If you compress a gas eventually its boiling point falls below that of the ambient temperature and it will condense back into a liquid. The latent heat of vaporisation is a measure of how much energy it takes to completely boil a mass of liquid to form a gas. This is usually many times more than the specific heat capacity. For example the specific heat of water is 4,190J/°C/kg, but the latent heat of vaporisation for water is in excess of 2,000,000J/°C/kg.

If you take a very cold liquid it can take on heat from its surrounding environment. If that level of heating is sufficient, and the liquid has a boiling point below the ambient temperature, it will boil the liquid to form a gas and in the process absorb a quantity of heat equivalent to the heat of vaporisation. If you take this gas and rapidly compress it to a high enough pressure it will condense to once again become a liquid. This liquid now has far more heat energy than it requires to sustain its liquid state and it dumps this excess of energy as heat. Using these two processes, boiling (or *evaporation*) and *condensation*, a heat pump can take low grade heat and pump it to produce high grade heat.

The efficiency of a heat pump, the coefficient of performance (COP), is calculated as the quantity of energy removed from the evaporator divided by the energy put into the compressor†. Well designed heat pumps have a COP of around 2.5 or higher. That is, for every 1J of energy supplied to the compressor 2.5J of low grade heat will be removed from the evaporator. So in total 3.5J of high grade heat (2.5J from the evaporator, 1J from the compressor) will be dumped into the condenser for each 1J of energy supplied to the compressor.

†   The efficiency is calculated as the energy saving over the total energy, equivalent to COP ÷ (COP + 1). At a COP of 2.5, for an energy input of 1 unit the energy output would by 3.5, the other 2.5 coming from the evaporator of the heat pump – in the process saving 2.5 units of energy. The cut in energy demand is the energy saving divided by the energy output, [(2.5 ÷ 3.5) × 100% =] 71.4%. The COP factor also works for the cost of running the heat pump, so it you put £100 of energy in, with a COP of 2.5, you would get free energy worth [2.5 × £100 =] £250 out.

The efficiency of the heat pump system depends a lot on the refrigerant. Ideally it should have a very high latent heat of vaporisation and a very dense gas phase. These properties ensure that the maximum amount of heat can be transported with the minimum amount of work being done by the compressor. *Chlorofluorocarbons* (or CFCs) were developed as refrigerants in the

1930s, and were used in most heat pump systems until the early 1990s when they were banned because of their harmful effects on the ozone layer. Today CFCs have been replaced with *hydrochlorofluorocarbons* (HCFCs). These are less damaging to the environment than the older CFCs but they are still based upon chemicals which can create problems in the longer-term – chlorine and fluorine. The other common refrigerants are gases such as propane or butane, ammonia, and even alcohol. These can be argued to be less harmful to the environment, although they don't have the same physical qualities as HCFCs.

The main problem with heat pumps is that they can't tolerate wide swings in the level of heat applied to the evaporator. That's because the compressor can only pump so much gas, and in order to do this the refrigerant must boil as a certain rate. If too much refrigerant boils it would affect the efficiency of the compressor. For this reason heat pumps are designed for a specific operating environment. So, for example, you couldn't cool a solar collector efficiently with a heat pump because it undergoes very wide temperature swings. This means that the most common source of low grade heat is the environment. For example, a heat pump can extract energy from a large body of water, such as a river or lake, or more commonly from a long loop of piping buried one to two metres below ground level. The heat can then be used for space heating or, like thermal solar, preheating water.

Most active renewable energy systems, like thermal solar or heat pumps, use electricity. We have to generate this electricity somehow or they will not function. The early use of renewable energy didn't require electricity generation because the motion created by a wind turbine or the water wheel was used directly – for example in the early textile mills or water pumps. What electricity gives us is a means of carrying the kinetic energy harvested from renewable sources and putting it to other uses. This is an important distinction. We do not generate electricity for the sake of creating electricity. Instead we should see it as a means of carrying energy from the environment and powering other processes that cannot harvest this energy directly.

Wind power is the densest source of renewable energy, and certainly the most high profile of all the methods of generating energy from renewable sources. The problem with wind energy is that it doesn't produce a constant power output. Some describe it as an *intermittent* source of energy. This is a misnomer because wind power it is actually *unpredictable*. It's not just the period of power production that varies, but also the power output. We can't predict the weather more than a few days ahead, and even then there are uncertainties about the precise wind speed at any time, and for this reason the power output.

The blades of a wind turbine sweep out a circular area as they rotate. The amount of energy extracted from the wind is proportional to the square of

Box 24. Producing Energy from Wind Turbines

Although we don't ordinarily perceive it, air is heavy. A cubic metre of air weighs about 1.2 kilos, so when air moves quickly we can get a lot of energy out of it. The level of energy can be calculated according to the density of the air, and the kinetic energy it contains at different wind speeds.

Wind turbines only get 15% to 25% of the theoretical energy available from the wind. Even so, the reason that wind power has been developed more quickly than the other potential sources of renewable energy is that it is one of the densest sources, and can produce a large amount of energy as electricity. The size of wind turbines has consistently increased over the last thirty years in order to achieve an ever higher level of power output from each turbine.

Even so, to produce the level of energy that we use today in the UK would require tens of thousands of turbines. Figure 40 shows the power output from the largest (at present) three mega-Watt (3MW) wind turbines if we were to cover a certain proportion of the UK with them. To illustrate the level of energy production, the upper line shows the percentage of the UK's electricity supply (in 2003) a certain area of land would produce, and the lower line shows the percentage of the UK's final energy supply (in 2003) a certain area of land would produce. The right-hand scale shows how many wind turbines are required to cover a certain proportion of the UK in 3MW wind turbines (based on a separation between the turbine of about 425 metres).

For example, covering almost 4% of the UK's land area (about 920,000 hectares) would producing 100% of the UK's electricity supply, and covering 4% of the UK would require just over 50,000 wind turbines.

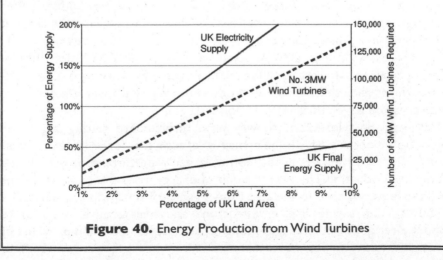

Figure 40. Energy Production from Wind Turbines

the blade length, but the cube of the wind speed. This is significant because it means that if we make a turbine twice as big, we get four times the power output. Alternately if the wind speed doubles, the power output goes up by a factor of eight. This can also be a problem because at very high wind speeds large amounts of energy are generated, and this could damage the turbine. For this reason turbines use different methods to control the maximum amount of energy that they produce during periods of high wind speed. The speed at which the blades rotate is proportional to the number of blades the turbine has. The more blades, the slower it will rotate. How the blades are designed is a critical part of improving the efficiency of the turbine. The profile and the length of the blades is designed to ensure that the speed at which the tip of the blade moves, compared to the wind speed, creates the ideal speed of rotation to extract as much energy as possible.

As an average power output, called the *load factor*, over the course of a year we may get between 30% (for turbines based on land) and 40% (for offshore turbines, based out at sea or on the coast) of the rated capacity of a wind turbine. So over the course of a year a large 3MW wind turbine will on average, at a load factor of 0.33 (33%), produce a continuous power output equivalent to a 1MW power plant. There are various options that might solve this, such as coupling a wind farm to a flow battery (an electrical storage cell, with a capacity of a few hundred giga-Joules, created by reacting electrolyte chemicals in vats the size of houses), but this would significantly increase the price of wind energy.

It would seem logical that larger wind turbines would produce a proportionately larger amount of energy. Today the largest production-line wind turbines are rated at around three mega-Watts, and some manufacturers have five mega-Watt models under development. However the concept that larger turbines produce more power is only true for a *single wind turbine*. When we put groups of wind turbines together the result changes, and the result is that larger wind turbines can produce less energy per unit of area than smaller wind turbines [Mobbs, 2004b]. Based on this data, five 2,300kW turbines would occupy the same area as fifteen 850kW turbines (which are 43% shorter), but the smaller turbines would produce more power over the course of a year than the larger turbines (see Box 25).

This begs the question as to why wind turbines are getting larger and larger. The answer is simple – the *economies of scale* that fewer, larger turbines create. What is driving the production of larger wind turbines is economics, not power production. The cost of a unit of energy from a large wind turbine is slightly less compared to the cost from a group of small wind turbines. This is because the engineering, infrastructure and maintenance required to support a certain amount of generating capacity costs less when provided by

Box 25. How Big is Enough? – *Economics vs. Efficiency*

Figure 41 is a scatter-graph of the height and capacity of 23 different wind turbines. The black triangles show how the hub height is related to the power output. This trend is illustrated by the solid line, and the slope of this line of the solid line confirms that larger wind turbines produce more energy.

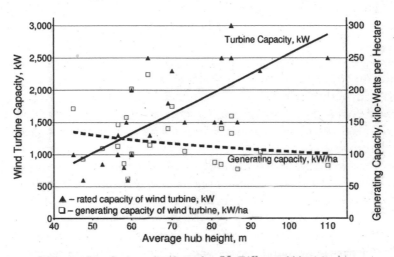

Figure 41. Output Analysis for 23 Different Wind Turbines
[Source: Mobbs, 2004b]

The right-hand scale shows the generating capacity for a group of identical turbines in kilo-Watts per hectare (kW/ha). The squares represent the power output per hectare for a group of identical wind turbines, weighted to take account of the difference in height of the turbines. The trend is shown by the dashed line, and illustrates that as the turbines get larger so the power generation per unit area falls.

The reason for this reduction in power output is simple. Bigger turbines are spaced further apart, because their hub height is greater. Therefore although they produce more power per turbine the output is less than a group of smaller turbines occupying the same area of land. The implication of this result is that a larger number of very small wind turbines occupying an area of land will produce more power than a few very large wind turbines.

† To calculate the density of turbines take 100 and divide by the average separation distance in metres, and then square the result to produce the number of turbines per hectare. For a 2,500kW, 85 metre hub height wind turbine the density, at five hub spacings (425m), will be (100m ÷ 425m)^2 = 0.055 turbines/ha, and the power output will be 0.055 x 2,500 = 138kW/ha.

larger turbines than smaller turbines. For the wind farm developer this means that the return on investment, or profit, on a few large wind turbines is greater than the return from many small wind turbines.

At the moment wind turbines are mostly backed-up by conventional sources of energy – so when the wind drops, the coal- and gas-fired power stations must work harder. What this means is that, in terms of the UK's total energy use, wind turbines don't produce a lot of energy. The mismatch between the location of wind power production and the power sources used to back it up also puts a strain on the national grid. Without some serious re-engineering of the national grid the practical level of wind power capacity is limited to between 10% and 20% of electricity demand – any more and the national grid would collapse.

When the wind blows across water it creates waves – an effect that can be amplified if the wind blows against strong currents in the water. The UK and Ireland, because of their location on the edge of the Atlantic ocean, are ideally placed to exploit wave power. There are four or five different mechanisms which can be used for extracting energy from waves, but the three most popular systems under development are:

- *Oscillating wave columns* – These are like a large bottle, open at the base, that stand in the open sea or on the shore line. As waves pass the water level inside the bottle rises and falls in response to the change in water levels outside. Like a piston in a cylinder the changing water level inside pushes and sucks air through the top of the column, turning a small turbine to generate electricity.
- *Articulated barrages* – These are floating ducks or snakes that twist and flex as a wave goes past. Snakes meet the wave head on, and the vertical buckling motion produces power. Ducks take the wave side on, creating a rotational motion that produces power. The mechanical action of the articulated floats compresses a fluid which, in addition to the pressure created by the other parts of the barrage, turns a turbine to generate electricity.
- *Surge devices* – these use a tapered channel to direct the waves up a slope, causing the energy in the waves to lift the water and fill a small reservoir above sea level. The energy of the water draining from the higher water level back into the sea turns a turbine to produce power.

The Royal Commission on Environmental Pollution rated wave power as the second largest source of renewable power after offshore wind. This is because water is very dense, and so waves at sea carry a lot of energy. However, unlike wind turbines, the fact that wave power doesn't scale-up to produce large capacity production sites means that research on developing viable wave systems has been fairly slow. However, over the next fifteen to twenty years, it could be developed as a viable option for powering coastal areas.

Another option for getting power from a turbine is to immerse it in the sea and harvest energy from the difference in water level created by the tide. Tidal power systems (put simply, dams that hold back sea water) generate energy because the gravitational force of the Moon lifts water on the Earth, giving it gravitational potential energy. It's the difference in water levels, between the inside and outside of the dam, before and after high tide that creates the flow of water required to extract energy from it.

The early forms of tidal power required that large estuaries were dammed up with a tidal barrage so that turbines could be turned by the difference in water levels before and after high tide. This is a very expensive option but it does generate a lot of energy, for many decades, once it is built. For example, the Royal Commission on Environmental Pollution highlight the proposed barrage for the Severn Estuary. This would have a capacity in excess of 8GW (that's equivalent to five or six conventional power stations) although because of the variation in the tide the output would not be constant. This means that the average power output for the barrage would be about the same as single large power station such as Drax (coal-fired) or Sizewell B (nuclear). At least sixteen other sites have been identified in the UK where tidal barrages could be built. However the ecological disruption to estuaries that tidal barrages create would disrupt some of the richest wildlife sites in the UK. For this reason tidal barrages must be considered a highly damaging source of renewable energy.

An enhancement of the estuary barrage idea is to construct a circular tidal impound in shallow water out at sea, from the sea bed to above the high tide level. Like a barrage this would generate power with the ebb and flow of the tide, but unlike a barrage it would not cause large-scale disruption to the aquatic environment. Although smaller, building an impound at sea is still a large undertaking. The problem with an impound dam it that it has to be very large to produce a useful amount of energy. Consequently the cost of the electricity produced would be higher. However, by building the impound dam with a number of small reservoir 'cells' it is possible to phase the movement of water in and out of the impound to create a near constant flow of energy. A scheme using this principle has been proposed off the north Wales coast, and more recently a scheme has been proposed for Swansea Bay [Tidal Electric, 2004].

The alternative to an impound system is to put a small water turbine into a fast running tidal current or *tidal stream*. In the same way that wind turbines turn in the wind a tidal turbine will turn in the water, but more slowly because of the lower speed of water compared to wind. Sea water is about eight hundred and sixty times more dense than air, so theoretically you would get eight hundred and sixty times the power output, or eight-six times the power output from a turbine one tenth the size (although in practice it's a lot less).

In the UK the first experiments with tidal stream generators were carried out in 1994, using a 10 kilo-Watt, 3.9 metre diameter turbine sited in the Corran Narrows near Fort William. A larger 300 kilo-Watt device was tested off Lynmouth in Devon during 2003. The best example to date has been a 20 metre diameter turbine in the Kvalsund Channel, near Hammerfest in Norway. This produces 300 kilowatts of electrical power and supplies 30 local homes. That's around 500 times the instantaneous power output compared to a wind turbine of the same size.

Another advantage of both tidal impound and tidal stream systems over sources like wind and solar is that the falls in output at low tide are entirely predictable – in the same way that the tides themselves are predictable day-to-day. This makes it far easier to factor in other sources to replace the lull in output from the tidal systems. Also, if you take into account the difference in load factors (0.4 for offshore wind, and around 0.65 for tidal stream turbines) then the difference in output between tidal streams and wind is actually over 800 times more for the same sized turbine.

However the main difference between wind and tidal turbines is cost. Tidal turbines are heavier and more complex since they must operate in sea water, and they are serviced using ships. Wind turbines, on land, can be serviced by a few people in a four-wheel drive van. Despite these difficulties, tidal streams could be a valuable energy resource. The figures in the Royal Commission on Environmental Pollution's report quote the potential tidal stream generating capacity in the UK at 4GW.

Hydro power, producing energy from flowing water, has the same benefits as tidal streams. Providing you can get a big enough flow of water, the density of water means that you can produce a lot of energy from a small device. The heat from the Sun drives the water cycle. This lifts water from the oceans, giving it gravitational potential energy, and dumps it down at the top of mountain ranges in the form of rain. As it falls back to the sea in streams and rivers this water can be used to generate power.

The traditional view of water or *hydro power* is the water wheel. For centuries water power provided energy to small-scale operations such as flour mills, and it expanding rapidly as an energy source at the beginning of the Industrial Revolution – until steam power displaced it as the main source of power for industry. However because of the inefficiencies of the slow-turning water wheel today most hydro power projects used specially designed water turbines that can extract three or four times the amount of energy from the flowing water compared to traditional water wheels.

Today there are many large scale hydro power projects being planned or developed around the world, generating mega-Watts of power from major rivers in upland areas. The problem with these schemes is that they also flood large areas of land, displacing perhaps tens of thousands of people. Consequently 'mega-hydro' has got a pretty bad press. A more useful alternative to the damaging mega-hydro systems would be the development of *micro-hydro* (or μ-hydro) schemes. Rather than a large impound dam to supply a large hydro plant, micro-hydro systems use only small quantities of water. Consequently they can be sited on small streams in upland areas and provide power to dwellings and small businesses in the immediate area. However, because micro-hydro generates a few kilo-Watts rather than mega-Watts, like domestic thermal solar energy the national renewable energy strategies tend to ignore the potential for micro-hydro.

In a micro-hydro system water is brought from the small dam or penstock down hill in a pipe. This creates a head of water pressure which is used to turn a special low-power turbine called a *pelton wheel*. This is made up of twenty or thirty small spoon-like blades fixed to a central hub. The spoons deflect the incoming jet of water slightly more than 90°, making it turn back on itself, and extracting a larger amount of energy than was possible from the old-style water wheels. They are also much smaller. The pelton wheel and generator housing for a small μ-hydro system that produces one or two kilowatts continuously is roughly the same size as a domestic washing machine. By connecting the hub of the pelton wheel to a generator the pelton wheel

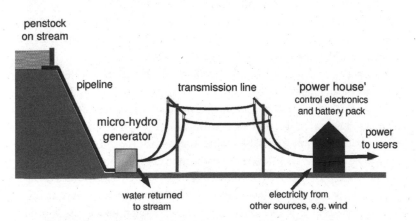

Figure 42. Diagram of a Typical Micro-Hydro Power System

Box 26. How Photovoltaic (PV) Cells Work

PV cells use semiconductors that trap photons from sunlight, turning them into electricity – hence, *photo-voltaic*. A PV cell is made from layers of silicon with a slightly different chemical structure, created by chemical impurities. The layers of the cell are translucent, allowing light to enter the whole cell.

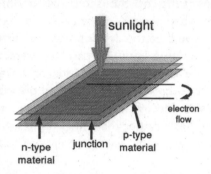

Figure 43. A Photovoltaic (PV) Cell

When light radiation of the required wavelength interacts with the atomic structure in the cell it displaces electrons generating an imbalance of charge across the cell. Whilst some of this charge is mopped-up within the cell and lost as heat, the presence of the junction between the two materials prevents the charge migrating across the cell. Instead it travels the long way round, via the wire contacts that are fixed to the n-type and p-type material, creating an electric current in the wires.

There are different types of PV cell. The main difference between each is their cost, and the amount of light that they convert into electricity, which affects their output current:

- *Monocrystalline* – amongst the most efficient PV cell (around 15% of the sunlight entering the cell is converted to electricity), although more expensive because it is engineered from a layer of pure silicon cut from a high quality silicon ingot (distinctive because of its blue appearance with regular thin wires running across the surface).
- *Poly- or multi-crystalline* – slightly lower quality PV cell, with an efficiency of around 12%, produced from recrystallised silicon ingots (distinctive because of its crystalline shimmering surface).
- *Thick-film* – small particles of crystalline silicon deposited on a substrate to form a thin layer.
- *Thin-film, triple junction and amorphous silicon* – advanced forms of semi-conducting material made of a layer of deposited silicon atoms rather than crystals. Some types, like *triple junction*, are more efficient at capturing lower light levels, and produce a higher energy output at higher operating temperatures compared to other types of PV cells. However, these types of PV cell have a lower operating efficiency, some as low as 6%, but are available at a lower cost because of simpler, less energy intensive production processes.

can produce a constant supply of electrical power. The power produced is likely to be lower than the peak demand by the users of that power, but because the micro-hydro system can develop a few kilowatts of power constantly for twenty-four hours a day it can charge batteries so that the large peaks in demand can be supplied as required.

Finally, it's possible to produce electricity from the Sun using *photovoltaic panels* – or *PV cells*. Although there are different types of PV cell available they all work the same way. The major problem with PV systems are their cost. For most small scale applications the cost of PV cells is around £2.50 to £15 *per Watt of capacity*. So whilst PV technology is one of the simplest renewable electrical power systems, it is also one of the most expensive. Other non-silicon technologies are currently being developed which are also, theoretically, less problematic to manufacture at the industrial scale. This could reduce the cost of PV cells by 50% over the next decade [ETSU, 2001], but even so this would still make PV very expensive compared to small-scale wind, thermal solar or micro-hydro system.

Whilst the early cells worked with a single semiconductor junction (see Box 26), today cells are being developed that absorb energy from a much wider spectrum of light by using more then one junction – for example, *triple junction* cells. The benefit of this is that whilst they may have a slightly lower maximum power output compared to the most efficient mono-crystalline cells, they produce comparatively more power at lower light levels. This is useful at higher latitudes because of the lower angle of the sun, and it increases the power output on cloudy days. Another curious fact about PV cells, and the reason why they are so inefficient at producing power, is that they lose a lot of the energy they capture as heat. When working at full power they get very hot, meaning that large PV cells are not really suitable for use inside a glazed building or enclosed space.

To produce a sufficient level of power the small PV cells are connected together in an *array*. To ensure that the power from the PV array doesn't overload the battery PV systems use a *charge controller*. This monitors the voltage level of the system and if it rises too high, when the battery has reached its full charge, it disconnects the panels. The battery pack that forms the core of the system usually uses *deep-cycle lead acid batteries*, specifically designed for use with energy systems. This means that you can size the PV array to produce the power that you need to use over the course of a day, and store it in the battery for use after dark. In practice you would only use 10% to 15% of the battery's capacity for storing your daily needs. This would mean that if you had a few cloudy days in succession you could slowly discharge the battery pack to continue providing power, and hope that the sky brightened up after a few days to begin the recharge cycle.

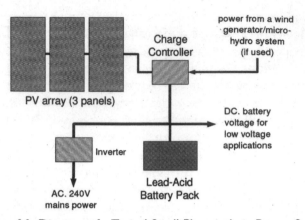

Figure 44. Diagram of a Typical Small Photovoltaic Power System

Perhaps the most effective, small-scale power system is when a PV array is combined with a small wind turbine of one or two metres in diameter. Mixing wind, PV, and perhaps micro-hydro, produces a *hybrid energy system*. The benefit of hybrid systems is that by using different small-scale energy technologies the natural variations in renewable energy supply, and the variations in human energy demand, can be evened out. The wind tends to blow when the sun doesn't shine, and vice versa. In this way, if the PV array isn't producing so much power, the wind turbine could make up the difference to ensure that the battery pack remained fully charged. Likewise there tends to be more wind and more rain in Winter, but less sunshine, when our energy demand is at its highest. As small scale wind and micro-hydro are more dense sources of energy they are also able to provide more power in the Winter when more energy is used.

Lead–acid batteries produce direct current (DC) at low voltages. This creates a problem as most household appliances are designed to use alternating current (AC) at high voltages. So when using a wind, PV, or a hybrid electrical power system you have two options. You could develop systems that run at the battery voltage using DC current. For example, a popular option is to convert the lighting in a house to use low voltage DC bulbs. The problem is that most modern appliances require mains power, and for this reason you have to use the second option – an *inverter*. This takes the DC current and oscillates it at a higher voltage to produce AC mains current.

Hopefully this review of renewable energy systems has provided a brief view as to what is possible with current technology. If energy supplies do become more unstable then increasing the role of renewable energy sources, particularly small-scale renewables, will be one solution by which we can gain energy security.

9. How Much Do We Need To Cut?

The peak in oil and gas supply over the next thirty or so years will drive home a stark lesson – *energy efficiency is meaningless in the face of actual energy shortages.* You can only reduce energy use if you have the energy to use in the first place. Choosing what uses of energy to cut or reduce is not the same as using energy more efficiently for the continuation of an existing use.

Energy efficiency has been a policy of British governments for the last thirty years. Remember the *Save It!* campaign of the 1970s, or the more recent *Home Energy Efficiency* or *Are You Doing You Bit?* campaigns? Whilst these efforts may have slowed the growth in energy use they have not actually reduced energy use. Energy efficiency is largely a voluntary measure, and it presumes that the public will choose to reduce energy consumption for either altruistic or cost-saving reasons. However, as we move past Peak Oil first price rises, and then actual oil shortages, will reduce the options to effectively one – what uses of oil can we cut or do without? This presents a wholly different motivation for planning reductions in energy use.

Figure 24, presented as part of a projection of how the peak in oil and gas supply would affect our energy supply, is rather abstract. It presents energy supply in terms of an overall figure for energy supply. Figure 45 presents the information used to create figure 24 in terms of our current energy use. It expresses as a percentage how much more (+) or less (–) will the energy we have available in the future be different to today.

It's likely that the gradual decline of oil supplies, coupled with higher energy prices generally, will curb consumption. So real-terms decreases in the UK energy supply, compared to today, are not likely until around the time of the peak in gas supply. But, as outlined earlier in the book, rather like the question of *'when will the oil run out?'*, this is not necessarily the issue of concern. As the energy supply stagnates rather than grows, it is prices that will affect people's lives well before they have to deal with having to physically cut energy use.

If energy costs more, and in turn the commodities we buy cost more, the immediate response is to make do with less. The previous three chapters have outlined how we can produce energy, but how much can we save?

The immediate problem we face in addressing the potential shortfall in energy is one of time-scales. Modern society has been slowly building and rebuilding itself for thousands of years. There are roughly 25 million households in the UK. Let's assume that we need to upgrade all of those households to be more energy efficient. That process can't be done quickly, quite simply because manufacturing the appropriate materials and devices takes time. So

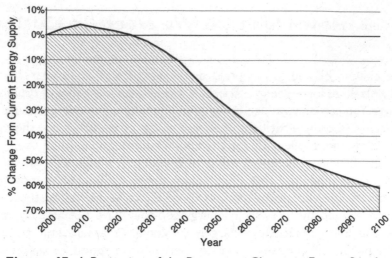

Figure 45. A Projection of the Percentage Change in Energy Supply

let's assume that we can re-fit 350,000 of those 25 million homes every year. At this rate it would take seventy-one years to complete the programme – arguably that's a little late as energy shortages are likely to happen, as shown in figure 45, within forty to fifty years.

Houses do use a lot of energy, but most of that energy used is utilised as heat – either to heat the air spaces in our homes, or to heat hot water. This has important implications for how we can source energy efficiently to supply domestic needs. Reducing the heat lost from the roof, walls and windows of a building could save a lot of energy. A large part of the housing stock in the UK is more than thirty or forty years old and these are difficult to upgrade to a highly efficient state without wholesale renovation – often involving the partial demolition of the building. So whilst lowering the heat loss from buildings can save a lot of energy, it is not a quick and easy solution for a large number of dwellings that, because of their age, are difficult to insulate without significant building work.

If you can't save energy by reducing heat losses, the alternative is to off-set energy demand by generating it at the point of use. The average dwelling uses 80GJ of energy per year, of which heating loads account for around 60GJ (75%) [DTI, 2002b]. A thermal solar collector, depending on its efficiency, may produce around $3.5GJ/m^2$ to $5.0GJ/m^2$ of heat per year [Jesch, 1981]. So a reasonably sized ($8m^2$) and relatively low tech. solar collector on the roof of a house might produce 28GJ to 40GJ of heat per year – over one-half of the annual 60GJ heat load.

Through the use of small-scale energy generation techniques in older buildings, where it would be difficult to easily reduce the fabric losses, the level of energy use can be reduced significantly without requiring wholesale demolition and redevelopment (which has its own energy implications when carried out on a large scale). To achieve this the public and policy-makers must adapt their concept of energy efficiency, instead focussing on managing the supply and use of energy. Putting greater emphasis on the use of active systems that efficiently conserve and recycle heat (such as heat pumps), and developing the small-scale production of energy (for example, thermal solar) to off-set the energy from external supplies.

The greatest problems will arise if abrupt climate change occurs in northern Europe. If you look at the houses remaining from the Eighteenth Century most of them have thick walls (usually described as *thermal mass*) and small windows as a measure to reduce heat loss during the Little Ice Age. It isn't just that thick walls reduce the heat loss. The use of thermal mass within a building works passively, buffering the temperature level within the living areas to maintain the internal temperature during Winter days. It also keeps the inside of the building cool in the Summer without the need for air conditioning – a useful feature if climate change were to make the UK hotter rather than colder.

The majority of the UK's housing stock, built from the late Nineteenth Century onward, are not designed in a manner that will minimise the fabric losses that will occur at the very low Winter temperatures abrupt climate change might produce. So a programme to renew or upgrade the oldest and poorest quality of the UK's housing stock is a high priority if we are to reduce energy consumption and ensure that these buildings are comfortable to live in – either because climate change makes the environment hotter, or because abrupt climate change makes it colder.

How fast such a programme could be carried out is limited by the capacity of the building industry – not just in terms of the building contractors, but the capacity of other companies to manufacture the materials that go into buildings. Let's assume that the UK building industry could build or renovate 350,000 dwellings per year. Renovating the UKs 25 million households would require 71.4 years to complete. So the energy benefits of the programme might take 20 years to show a significant return.

If we assume that within the whole domestic sector:

- The energy used for space heating, which makes up around 50% of the energy use in the domestic sector, were cut by 50% through reducing thermal losses and another 10% by using some passive solar heat, then energy use per household would fall

Box 27. The Problems with Poorly-Conceived Energy Schemes

The problem to date is that successive governments have not sought to implement schemes that would change the way the domestic sector uses energy. All the energy efficiency schemes to date have resulted in almost no change in the intensity of energy use in the domestic sector [DTI, 2002b].

A more recent problem is that the UK government has become fixated by certain forms of media-friendly renewable energy (such as photovoltaic cells or wind) rather than looking at the energy issue as a whole. For this reason many of the schemes to reduce home energy use have been poorly conceived and implemented.

For example, in 2002, the government announced grants to encourage the development of solar electricity systems for homes [DTI-EST, 2002] with a £20 million grant to the Energy Saving Trusts (who develop energy efficiency projects on behalf of the Department of Trade and Industry). However, although the publicity was aimed at homes, when you read the detail it is clear that companies or institutions might benefit preferentially from the scheme [DTI-EST, 2003].

The *Get Solar PV* scheme encouraged people to produce a proportion of their electricity consumption using photovoltaic panels. But the basis of the scheme, if we wanted the best result in energy terms for the £20 million that was spent, is clearly wrong. PV cells are not the best way to achieve large energy savings in a short period of time. Compared to thermal solar collectors, PV cells collect just a small fraction, perhaps 10% or less, of the energy radiated by the Sun. Also most of the energy people use in their homes is heat, not electricity, and so the transformation of solar energy directly to heat is more efficient than using an intermediary form of energy such as electricity.

For a similar energy output PV systems also cost at least four times more to install than a thermal solar system. Overall, for the amount of energy produced with the £20 million worth of PV schemes, perhaps 38 times more energy[†] could have been produced had these grants been aimed at thermal solar systems.

Perhaps due to these criticisms, the various grants for renewable energy are now being reviewed by the government, and funding might not be renewed for the *Solar PV* scheme in the future [Guardian, 2004r].

---

† This factor holds true irrespective of the size of the grant. If one unit of PV energy is produced by £1 of expenditure, but thermal solar costs one-quarter (0.25) the cost PV-based energy and produces nine or ten times the energy output (let's say 9.5), then that £1 would have produced [(1 unit x 9.5) / 0.25 =] 38 times more energy if it had bought thermal solar technology.

24GJ per household per year. The space heating load could be reduced further if passive solar design were used to trap heat in the building fabric to maintain the temperature of the house into the evening hours.

- The energy used for water heating, which makes up around 25% of the energy use in the domestic sector, were cut by 50% using thermal solar systems then energy use would fall 10GJ per household per year – perhaps slightly more if heat pumps were used to recover waste heat to pre-heat the hot water system.

Each household, after such a redevelopment or renovation, would save around 34GJ/year – or 42% of current domestic energy use. At 350,000 improved households per year, this would reduce energy consumption by 12PJ each year, or a total of 857PJ each year when the renovation programme was complete. Assuming that other efficiency factors (pessimistically) were to create savings of around 10% over this period, after seventy years energy consumption in the domestic sector would reduce by nearly 1,000PJ/year, which is about half the current level of energy consumption in the domestic sector.

These levels of energy reduction in the housing sector are not theoretical [BBC, 2004j]. New low energy housing systems [RCEP, 2000] available today, which use standard building techniques, consume 40% less energy than homes built to current (2004) building regulations. These additional energy saving measures add less than one percent to the cost of a new home, so we should question why the government will not implement these standards through the building regulations system. Even though the government have made noises about requiring that homes produce energy using solar [Observer, 2004c], in reality the changes to building regulations this would require are some years from being completed.

Other more experimental designs, such as the Beddington Zero Energy Development in London [BedZed, 2004], aim to reduce energy consumption to 10% of the level used in the average home today. However, these types of development are not likely to be phased in for perhaps ten or twenty years as the systems they use are tested, and eventually enter the mainstream of construction industry practice.

One side-effect of these changes would be that the demand for energy and/or generating capacity is significantly reduced in Summer because of the use of solar heat. This means that although consumption from external energy sources would be lower overall because of the reduced thermal losses, most of the energy demand will be during Winter months. If the UK switched to using more biomass this cycle of energy use would be an advantage. This is because the biomass would become available at the time energy demand would begin to increase – in the Autumn, following the harvest – so the need to store biomass all year round would be reduced. Animals also produce more organic waste, suitable for anaerobic digestion, whilst they are kept in sheds during the Winter months.

Whilst housing may be a large sink of energy within the life of the average person, so is their consumption of food. Modern agriculture is highly energy intensive, but the majority of that energy isn't consumed by machinery. It's the fossil-fuel sourced fertilisers and pesticides, and the large amounts of energy used to heat glass houses or fuel the transport of food in cargo aircraft, that mean in energy terms we eat more calories of oil and gas in the average meal than the calorific value of the food itself.

In terms of actual food consumption, the 3.6GJ of food energy an average person in the USA consumes each year is created from 80GJ of biomass and 7GJ of agrochemicals. This produces 6.5GJ of raw food which is processed using 29GJ of energy during transport and processing [Weizsäcker, 1998]. Dividing the calorific value of the food by the energy used to bring it to your plate (the agrochemicals, transport and processing) that's an efficiency level of 10%. If we include the solar input that generates the biomass, the level of efficiency drops to 3%. However, this large energy input isn't reflected in the price of food as energy is still cheap. Even though half of the energy used in the whole food production system is sourced from fossil fuels, and 14.5GJ of the 29GJ used in processing is due to transport, these costs only represent 7.5% of the overall cost of the food [CRE, 2001].

The issue of energy use within food production and distribution is often analysed in terms of *food miles*. This is a measure of energy intensity in terms of how far food travels from the farm to the consumer's plate. There are many ways to assess the impact of food use on the economy and energy use:

- Department of Transport statistics show that 40% of all road freight transport is carrying food and food products [DETR, 1999]. Apart from being the biggest operation within the UK road freight industry, using the figures from 2001 [DTI, 2002b], food transport would have consumed 251PJ/year, or 11% of the energy used in the entire transport sector.
- The development of regional and global economic systems also means that for a variety of reasons food is being moved reciprocally, using cheap energy for the sake of economic gain. For example, in 1997, the UK imported 126 million litres (23 million gallons) of milk whilst at the same time exporting 270 million litres (50 million gallons) of milk [SAFE, 1998].
- Using similar methods to life-cycle assessment, it's possible to

add up the food miles within a single meal by adding together the individual miles for each of the raw ingredients. This produces a figure for an average meal, or for a period of consumption. Using this process studies have claimed that some processed meals contain 24,000 food miles. However, those who eat seasonally available foods, and who source most of their food locally as raw ingredients rather than as processed food might only eat 376 food miles in an average meal [Sustain, 2001], a saving of 98%.

There are a number of reasons why the distance food travels is increasing, but globalisation is only one factor. Perhaps perversely, the increasing demand for more low-intensity organic food is leading to food being sourced over greater distances because the UK has such a low level of organic production. Another factor is that the increasing concentration of ownership in the UK grocery sector, between just a few large supermarket chains, leads to a higher level of centralisation around a few large food distribution centres – meaning that food must travel further between producer, distribution centre and local superstore. Finally, the increasing trend beyond ordinary processed foods towards ready meals, again produced by a small number of large food processing companies, also leads to lengthier transport due to centralisation, and the sourcing of cheaper ingredients from across the globe. The consumption of ready meals rose 44% in the UK between 1998 and 2002, and it is likely to increase by another 25% by 2007, when half the market for ready-meals in Europe will be within the UK [BBC, 2004l].

One factor related to Peak Oil, and then at a later date to Peak Gas, is that the precariousness of energy supply over the next twenty to thirty years endangers our systems of industrial food production. It's arguable that since Britain urbanised in the late Eighteenth Century it has been increasingly reliant upon food delivered from other parts of the globe [BBC, 2003e]. If our industrial food supply or global transport systems failed it would cause great social and political problems. This is entirely possible if oil becomes scarce, or, as outlined recently by the US Pentagon's future strategy planners, because abrupt climate change disrupts our current processes of food production [GBN, 2003]. Other problems include the fact that modern crop species have been bred to utilise high levels of fertiliser, and so reducing energy input would require a wholesale revision of crop strains as well as the manner in which those crops are grown.

Energy costs, and eventually a lack of cheap fertilisers sourced from petrochemicals, mean that a large proportion of agriculture will need to switch back to traditional organic farming methods over the next thirty or forty years. How much energy this saves depends upon how agricultural practice changes to take account of the price of energy. It also depends on whether new food production systems develop other spin-offs (such as the digestion of animal wastes to produce energy) which would reduce the net energy use in food production.

If the transport requirement for food in the UK were to halve, as production switched to more localised systems, that would save 125PJ/year – 5.5% of the energy used in the transport sector. Add on the back of this the potential savings from the consequential global reduction in food miles, and the total savings could be a few hundred peta-joules per year. However, given that petroleum will be hit first, both by rising prices and then shortages as we pass Peak Oil, it's likely that the savings will be a lot greater because the increase in agricultural costs would price-in more localised food production systems.

The transport sector is the largest use of energy in the UK. Transport is also the largest user of petroleum products, and so will be most exposed to the rise in prices following the date of Peak Oil. Consequently, transport has to change. To date, although there have been some experiments in the transport sector, such as *hybrid cars* that burn petroleum fuels more efficiently to produce electricity to drive the car, many of these systems have failed to produce the intended saving in energy use [Wired, 2004].

Reduced the supply of oil may lead to a slight reduction in greenhouse gas emissions, but a large part of the reductions in carbon emissions planned by the UK government are already predicated upon *fuel switching* – changing the type of fuel that transport systems run on. It's likely that within the transport sector energy demand will switch from petroleum towards *liquefied petroleum gas* (LPG). Natural gas could also be turned into liquid fuels for the transport sector should LPG run short. Both of these options have already been identified as a matter of policy by the UK Department of Transport in their Powering Future Vehicles Strategy [DfT, 2002]. In the Department's recent update on progress it would appear that they are already seeking the conversion of many vehicles to LPG within a short time-scale in order to meet European emissions targets [DfT, 2003].

The problem is that the required volumes of natural gas might not be available at a reasonable price for the production of LPG – something that the Department for Transport's consultants didn't seem to consider in detail as they were primarily studying the impact on carbon emissions [Ricardo, 2002]. Manufacturing LPG from other hydrocarbons is as expensive as producing some of the most difficult secondary oil deposits – up to $10 to $16 per barrel of oil (about two to three times the current price range for oil production) [IEA, 2001]. If hydrogen were to develop as a fuel, perhaps using hydrocarbons like natural gas as a precursor fuel, that too will push up the demand for gas to substitute for the loss of petrochemicals.

It doesn't matter whether we opt for LPG or hydrogen to reduce carbon emissions, unless we actually cut the distance people travel in cars each year the proposed changes to our transport system are just storing up problems for the future. In reality using LPG, or making hydrogen from heavy oil deposits or gas, is really just playing with the order of the deck chairs on the Titanic. Biodiesel isn't a realistic option either because we don't have the available land area to produce that much fuel (see Box 20). Eventually the transport sector will have to realise the inevitable truth, enforced by the *First Law of Thermodynamics* – as less energy is available from hydrocarbon deposits, the amount of available energy will contract and the level of transport use will have to be cut significantly. Rather like improving the housing stock this isn't something that can just be done overnight. Over the last fifty years the UK has structured its development patterns around the use of the car. In fifty years time that will be a very expensive option. Therefore, over the next fifty years, we have to plan for real-terms cuts, year on year, in the use of private transport (see Box 28).

Of course, a large proportion of the energy we use in the UK is wasted, mostly as a result of electricity generation. As shown in figures 10 and 11, one way or another a third of the energy we use is lost from the system before it reaches the consumer. Renewable energy sources, producing electricity or heat directly, are useful in terms of their efficiency because they have a multiplier effect. As fossil fuel electricity is inefficiently created from fossil fuels, displacing one unit of fossil fuel electricity with renewable electricity saves between two and three units of fossil fuels. The problem is fossil-fuel based energy demand is growing faster than the development of renewables, and in any case renewables could not practically replace our current use of primary energy sources for power generation – we don't have the land to make that much power.

The easiest way to tackle the largest area of energy loss from the UK economy would be reform the greatest source of loss – our current system of large-scale power generation that feeds the national grid. Various organisations are now advocating the need to restructure the national grid. For example the Parliamentary Office of Science and Technology, in their report on electricity networks [POST, 2001], highlight the potential to develop one third of the UK's electrical power from embedded generation – generating capacity that forms part of local networks rather than providing solely for the grid.

A key feature of embedded generation is that it allows power storage to be integrated into the local grid. This would mean that the output from smaller renewable or low carbon energy sources could be buffered in a storage system overnight and then used to meet peaks in demand the next day. More

Box 28. Shifting Energy Demand in the Transport Sector

In 2001, road transport used 42 mtoe (1,758PJ) of energy per year [DTI, 2002b]. In 2003, cars, vans and taxis travelled 900 billion passenger kilometres (bpkm), using 26 mtoe (1,089PJ) of energy. Rail transport was 80bpkm using 1.5 mtoe (63PJ). And buses and motor cycles travelled 80 bpkm and 7 bpkm respectively, using 1 mtoe (42PJ) of energy between them.

Let's assume that we want to shift 30% of the demand for cars to rail, and another 30% to buses. How much energy would this save?

- 60% of the car distance travelled would be 540 billion passenger kilometres (moving 270bpkm to each of the other two modes), removing 653PJ of energy use per year from cars (327PJ/year to each of the other two modes).
- Shifting 270 bpkm to rail would add an additional 213PJ of energy demand, bringing the annual total (for 350 bpkm) for rail to 276PJ. The energy that would have been used for this distance of car-based travel is 327PJ/year, so the net energy saving shifting this travel demand to rail would be 114PJ/year†.
- Shifting 270bpkm per year to buses would add an additional 142PJ of energy, bringing the annual total (for 350bpkm) for buses to 184PJ. The energy that would have been used for this distance of car-based travel is 327PJ/year, so the net energy saving shifting this travel demand to buses would be 185PJ/year‡.

So, moving 60% of current car use to buses and trains creates a net energy saving of 299PJ per year. That's 17% of the 1,758PJ per year currently used in the road transport sector, and a saving of nearly 13% of the 2,390PJ per year used in the transport sector as a whole... *not a lot really!*

The fact that such as major shift in the way transport operates would produce such a small saving in energy use illustrates one of the key issues related to Peak Oil – we travel too far by mechanised transport. Efficiency savings alone are not going to significantly reduce energy use in the transport sector. Therefore the longer term aim has to be a real-terms reduction in the energy used/distance travelled by power transport. This can only be done through a greater localisation of the economy, so eliminating the need for people and goods to regularly move long distances.

† Train use would increase [270 billion ÷ 80 billion =] 3.38 times, increasing train energy demand by [3.38 x 63PJ =] 213PJ and consumption to [63PJ + 213PJ =] 276PJ per year. The net figure is 30% of the car transport energy, [1,089PJ x 0.3 =] 327PJ, less the rail transport increase, which is [327PJ/year – 213PJ/year =] 114PJ/year.

‡ Bus use would increase [270 billion ÷ 80 billion =] 3.38 times, increasing bus energy demand by [3.38 x 42PJ =] 142PJ, and consumption to [42PJ + 142PJ =] 184PJ/year. The net figure is 30% of the car transport energy, [1,089PJ x 0.3 =] 327PJ, less the bus transport increase, which is [327PJ/year – 142PJ/year =] 185PJ/year.

storage capacity, using large-scale energy storage devices such as flow batteries that can store giga-Joules of energy, would also provide power when unpredictable or intermittent sources, like wind and tidal power, were not producing energy. The overall effect is to reduce the amount of generating plant on standby because the peak demand will be met from the storage systems, not the power generation system.

The greatest impact of localisation of power generation is that it enables the development of combined heat and power systems – which would significantly increase the efficiency of power generation. A large proportion of the energy loss in power generation is a result of power stations dumping most the energy they use as heat into the environment. By generating power locally most of this wasted heat could be supplied to nearby buildings to displace energy that would have been used for heating. In turn, the increase in efficiency would create large savings in the primary energy used by the UK economy. For example, if one-third of the UKs coal- and gas-fired power generation capacity were transferred to localised production, with an increase in energy efficiency to 60% through the use of the waste heat for heating, it would save 27TWh (96PJ) of energy per year, 16% of the coal and gas energy used for power generation in the UK.

The issue of de-nationalising the national grid isn't directly related to energy reduction. It would also create opportunities for small wind, micro-hydro and even tidal power or wave power schemes to be developed to supplement local energy supplies. Biomass and anaerobic digestion would also fit better into such a system because they would work more efficiently, and so require less land, if they are used in a combined heat and power network. Together this would displace the use of primary energy sources such as coal and gas, and would significantly reduce carbon emissions.

The ideas for saving energy, outlined above, are very general. It is likely that more savings could be identified if a more detailed study of the whole UK energy system were carried out. However it is politics, not efficiency of energy use, that will be the major factor in how our future use of energy is planned. In the short-term the current trends and national policies for energy could create a lot of instability within the energy market, possibly leading to reductions or interruptions in supplies [NERA, 2002]. Such issues make headlines, and even BBC drama documentary programmes [BBC, 2004i]. The reality is that whilst we are dependent on the large-scale importation of energy resources we will always be subject to the economic pressures of the global energy market.

At the moment governments across the globe are carrying out strenuous rounds of negotiations to reduce carbon emissions and avert climate change. However, the flaw in the negotiations over carbon emission controls is that they don't actually restrict energy use and the emission of carbon – they just make the use of energy more expensive. If you have the money you have the potential to buy all the carbon emission credits you want, and you are able to keep burning fossil fuels.

The Peak Oil issue is something wholly different to the control of carbon emissions. Oil, and then gas, will just not be available in the quantities they were previously. To begin with they will become more expensive, and the impact of energy price rises may be as disruptive to modern society as the oil running out. The global economy is so centred on the burning of fossil fuels that changes in prices would create instability within the financial systems that run the world today. Eventually, after perhaps fifteen or twenty years of high prices, oil will become scarce. The UK's current energy policy is predicated wholly on the reduction of carbon emissions. Consequently it is not looking far enough ahead to provide a strategy for what foreseeable price rises and volume reductions in oil and gas supply will mean to our society.

How you interpret the data on the world's remaining oil and gas reserves will inform your view on how long they are likely to last. If you divide the proven oil and gas resources – 1,148 billion barrels of oil and 176 trillion cubic metres of natural gas [BP, 2004] – by the current level of use – 28.5 billion barrel per year for oil and 2.6 trillion cubic metres per year for natural gas – you will under-estimate the lifetime of the resource (for example, on the basis of current proven reserves, oil will last 40 years and gas 68 years). This under-estimate of the resource life occurs because the effect of the peak in production is to lower our consumption level, making the resource last longer.

However, if we project the peak in oil and gas production as half the total resource figure we would underestimate the dates of Peak Oil and Peak Gas. This is because as the rate of consumption increases the point at which the peak in production is likely to take place moves closer. Then, once the peak in production is reached, the remaining reserves will last longer because production volumes will have to fall.

So, if the peak in oil and gas supply reduces our energy supply, and we can only reduce our consumption of energy so much through greater efficiency measures, can we replace this lost energy from other sources?

The obvious answer to this shortfall in energy output would be, *what about the coal?* It is true that the UK does have large coal reserves, but the problem of climate change makes them unusable unless we trap the carbon emissions – using carbon sequestration. Even so, if we did use our own coal it wouldn't last long. There is a common myth that the UK has two-hundred years worth of coal buried beneath it. This may have been true in the 1950s, based upon 1950s levels of electricity consumption, but today we use so much energy that the life of our coal reserves is far shorter.

In 2002, the UK used 58,400,000 tonnes of coal, just under half of it imported [DTI, 2003a]. The 48,500,000 tonnes of coal burnt in power stations only produced 34% of the UK's electricity supply. If we were to produce all our electrical power from coal it would require nearly three times the current level of coal consumption. The UK's proven coal reserves, on BP's data [BP, 2004], are 1,500,000,000 tonnes. So at current rates of coal consumption the UK's coal reserves would only last 26 years, or less than ten years if we tried to get all our electricity from UK-sourced coal. Even then, burning coal would only deal with electricity consumption. A far larger quantity of energy is consumed as gas and oil directly.

The only long-term option for using coal within the UK, and in fact across Europe, would be importing large quantities from the world's major reserve holders – the USA, Russia and China. Currently the world uses around 4.5 billion tonnes of coal per year [IEA, 2001]. Taking all proven reserves together at the current rate of use would mean that there are 219 years of supply in reserve. However, the quality of coal, and its calorific value, varies. Whilst anthracite, steam coal or bituminous coal have a calorific value as high as 33GJ per tonne, and the lower quality UK power station coals around 26GJ per tonne, just under half of the world's reserves are brown coal and lignite which have a calorific value of 20GJ to 15GJ per tonne, sometimes less. So the coal reserves of the world are not equally as useful for power production, and a representative calorific value for the global reserve would be 20GJ to 21GJ per tonne.

If we scale-up this analysis to look at the world's energy supply, the 4.5 billion tonnes of coal the world consumes each year represents 26% of primary energy consumption [BP, 2004]. World coal reserves are just over 984 billion tonnes so at current rates of consumption the coal would last more than 200 years, but if we increase the proportion of the world's energy supply sourced from coal then the lifetime of coal resources shrinks significantly – to less than one hundred years if coal were to make up 60% of the world's energy use.

However, the use of coal would exacerbate the greatest problem facing the world today – climate change. The principal problem with burning coal is that you have to sequester the carbon produced by the process and store it. If not, the levels of greenhouse gases produced from burning the world's coal reserves would make the process of climate change run-away with itself, heating the world in excess of 10°C or 12°C. The last time that happened, 250 million years ago during the Permian extinction, most of the world's animal species died out. So, rather like the problem of radioactive waste disposal, if we predicated coal burning on the basis that carbon sequestration and safe disposal of the carbon waste had to be proven, this might restrict the use of the world's coal reserves. Consequently, coal wouldn't provide as much energy.

There is always the nuclear option. But as outlined earlier in the book, that is not viable unless you can get fast breeder reactors working to extend the life of the world's uranium reserve. The IEA's figure for known conventional uranium reserves (expressed as pure uranium metal) is 4 million tonnes, although the European Commission put it at 2 million tonnes in their energy green paper [EC, 2001]. Estimates of world uranium consumption vary from year to year – between 31,100 tonnes and 64,000 tonnes [IEA, 2001]. This provides 6.6% or the world's energy supply [UNDP, 2000], although more recent data from the BP survey puts this at 6.1% [BP, 2004], and it is likely to decline further at the reactors built in the 1960s and 1970s are decommissioned and replaced with non-nuclear sources.

If we assume that the world used 64,000 tonnes per year (roughly the current consumption), then the world's uranium reserves would last 63 years (on the IEA estimate of uranium reserves) to 31 years (EC estimate of reserves). If we increase the proportion of the world's energy supply from nuclear to around 20% or 25%, consistent with a switch to nuclear power in developed states, then uranium consumption would increase 3.7 times to around 237,000 tonnes per year. In this case the world's uranium reserves would only last 17 years (IEA) to 8 years (EC).

Finally, renewable energy sources could be developed as fossil fuel sources become scarce, but they won't supply the large amounts of energy the economies of Western Europe currently use.

Although the UK is increasing the amounts of energy it develops from renewable sources the amounts produced compared to the level of primary energy consumption are minute. The contribution from different renewable energy sources in 2003 are shown in figure 37. A problem with the Department of Trade and Industry's current definition of renewable is that some of the energy sources are not really renewable. In 2003, around 50% of the UK's renewable energy sources were reliant upon the DTI's questionable view of what constitutes renewable, and only 11% came from the renewable wind, solar or hydro power sources – the rest came from low carbon wastes and biofuels. As outlined earlier in the book incineration, or the use of landfill gas, rely on the continued production of large volumes of waste by society, and the inefficiency of energy use implicit in the process of waste production. This is something that cannot happen as we move to a truly sustainable energy policy.

As outlined in the chapters on low carbon and renewable energy, it is a forlorn hope to expect that we can develop a self-sufficient renewable energy economy in the UK. We just don't have the land area to create that much energy and to feed ourselves. The only way we could maximize production from renewable sources would be by developing an appropriate mix of sources, depending upon local circumstances, and in particular developing a large number of small-scale sources rather than large grid connected systems. What this is likely to produce is less than 40% of our current energy consumption. Therefore we must cut our use of energy to a level where we match our level of consumption to the level of production from renewable sources. Given that such a course of action would require the wholesale redesign of our society, and the way it operates, the key question about this solution would be, how long have we got to implement these changes?

Of course, there is an opposing view. Instead of cutting energy use we utilise every possible mineral energy resource – the coal and uranium included – irrespective of the consequences on the global environment. The rationale for this is that, on the cost-benefit analyses promoted by researchers such as Bjørn Lomborg, this would cost less in terms of loss of productivity from the global economy than restricting economic growth in order to lower fossil fuel use and the emission of greenhouse gases. If we did adopt this policy, ignoring the environmental implications, the key question would be, how long could the policy work for given that we know these energy sources are finite?

To answer the two extremes of possible future energy policy we need to produce two opposing models of energy use. We take the data presented throughout the book, on the availability of oil and gas as they pass the peak in production, and on the productivity of various other energy sources, and look at how energy supply might change over the next century or so.

The first model, let's call it the *Burn Everything Model* (see Box 29) – would use absolutely every mineral fuel source that we have available in order to keep our conventional, large-scale energy systems running. The aim of this would be to maintain the long-term increase in the UK's energy supply for as long as possible.

The second model, let's call it the *Reduce and Renewable Model* (see Box 30), is an antithesis to the first. Nuclear power would be abandoned in the very near future, and we would deliberately run down the use of mineral energy sources before they reach the peak in supply (to invest the higher costs of those fuels into other, renewable, energy sources). As far as possible we would offset the energy lost through the use of renewable energy sources, but the most important component in this process are real-terms reductions in certain aspects of our current energy use.

The important point about these two models is their end points. If we "burn everything", then in one hundred and twenty years time we are in exactly the same position as the reduction model – a primary energy supply of around 3EJ to 4EJ per year.

The key fact about the *Burn Everything Model* is that the reduction in the availability of oil and gas puts greater burdens on coal and nuclear power,

Box 29. The "Burn Everything Model"

The assumption in the *Burn Everything Model* is that the UK continues to use the same proportions of oil, gas, coal and uranium – as far as the available level of supply will allow – that we use today until those energy sources are exhausted. As oil and then gas reach a peak and then decline, it is increases in the UK's coal and nuclear energy capacity that keep the total energy supply growing at roughly the same rate as the past twenty years.

The key features of the model are:

- The amounts of oil and gas available are based upon recent studies of oil and gas supply [e.g. Laherrère, 2001]. The global daily supply predicted in these studies is apportioned according to the UK's current use of these sources –
  - 2.2% of the world's oil supply (218EJ, or 39 billion barrels over the next 120 years), and
  - 3.7% of the world's gas supply (333EJ, or 8 trillion cubic metres of natural gas over the next 120 years).
- The amounts of coal and uranium used are based upon the UK's current use of these sources –
  - 1.5% of the world's coal supply (271EJ, or 13 billion tonnes over the next 120 years), and
  - 3.3% of the world's nuclear energy (80EJ, or around 100,000 tonnes of uranium over the next 120 years) .
- The assumption is that globally the use of coal and uranium will be phased to keep energy supply increasing at roughly the current levels.
- Clearly it would be theoretically possible to increase the use of uranium and coal to keep the energy supply growing after 2040. However, that's not a practical reality, and so the availability of coal and uranium has been modelled as a normal distribution around the peak date.
- The peak in supply for each of the major fuels are –
  - Oil, 3.56EJ per year around 2015,
  - Gas, 4.82EJ per year around 2035,
  - Uranium, 1.92EJ per year around 2040, and
  - Coal, 2.96EJ per year around 2075.

The major uncertainty with this model is the availability of energy. The basis of this model is *business as usual* – that we continue to use energy as we do today. It assumes that we carry on using the same amount of energy in a world where countries will be competing for the same energy resources. Consequently, the model does not take into account the possible fall in energy use due to higher prices, or the increases in the energy that might be available due to the higher efficiency of use.

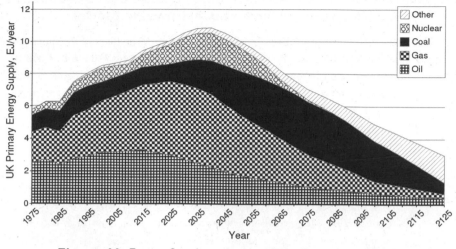

Figure 46. Energy Supply under the "Burn Everything Model"

exhausting the reserves faster. This means that whilst we might get another thirty to forty years of increasing energy supply, after this it becomes physically impossible to maintain these increases because the energy resources can no longer provide an appropriate supply.

The *Reduce and Renewable Model* shows an ever increasing level of reduction in energy supply throughout the middle of the century. The primary reason for this is that the 0.5% to 1% real-terms cut in energy use, year on year, compounds to accelerate the decreases in energy use. In terms of carbon emissions this would appear to cut emissions to zero within seventy years. However given the amount of fossil energy that is already in the biosphere that is going to be processed and recycled for the foreseeable future, especially plastics, fossil carbon emissions would still exist at least until the end of the century.

The other key difference between the end points of the two models is that by following the *Burn Everything Model* climate change is likely to have been made far, far worse. Across the world there will also be a few million tonnes of highly radioactive waste that would need careful management, but little energy to allow the management of the waste to be carried out. However, if we look at it from the opposite extreme, by following the *Reduce and Renewable Model*, in one hundred and twenty years time the citizens of the developed nations will be materially poorer, and less mobile. What's also important about the *Reduce and Renewable Model* is just how little energy, compared to today's level of energy use, the renewable energy sources actually produce. It should also be noted that the scale of the renewable

Box 30. The "Reduce and Renewable Model"

This model takes the opposite approach to the *Burn Everything Model* because it assumes that, in advance of the peak in oil and gas supply, we reduce our use of fossil fuels one or two percent per year. This could be as a result of higher prices driving down demand, or restrictions on the use of these fuels due to climate change.

The key feature of the model are:

- Nuclear power in the UK is phased out completely in the near future, and there is a limit put on the future use of coal.
- Instead of investing ever more money in energy supply the underlying trend in this model is that between 2010 and 2065 to 2070, the entire UK economy is regenerated in a way that reduces energy use and creates a greater reliance upon renewable energy sources.
- In the interim period the UK's coal generation plant are converted from standard combustion to IGCC in order to reduce carbon emissions and increase efficiency.
- The use of gas is switched from combustion to combined heat and power, and ultimately to more efficient forms of CHP, such as fuel cells which use natural gas as a precursor fuel.
- At the same time the national grid is made redundant, as more and more power generation switches to local energy sources and CHP systems, which creates a large reduction in the amount of primary energy wasted due to energy transformation.
- There is also a significant reduction in the use of natural resources, and large reductions in the use of cars, which create further real-terms savings in energy use.

The targets for renewable energy and low carbon energy sources are rather pessimistic. The targets are the same, or in some cases less than those cited by the Royal Commission on Environmental Pollution [RCEP, 2000]. This recognises the conflict between the need to live and produce food, and the need to produce energy, from the land area of the UK (this model abandons the idea of energy imports as a means of meeting energy demand). Therefore the total energy supply, after the use of oil, gas and coal ceases, is roughly one-quarter to one-third of the UK's primary energy supply.

In reality most of that energy isn't "missing". It's a result of the 40% reduction in use due to more efficient power generation and the reductions in private transport use. Therefore the actual "loss" of energy to the UK economy is about 25% to 30% of our current energy use, mostly made up of changes to the energy production, manufacturing and food supply systems.

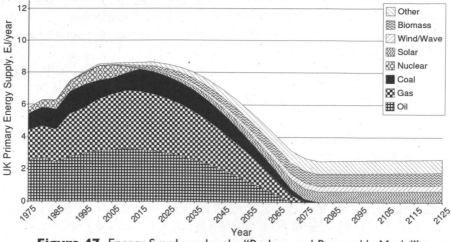

Figure 47. Energy Supply under the "Reduce and Renewable Model"

energy development in the *Reduce and Renewable Model*, even though pessimistic in terms of what is theoretically possible, is still way beyond the scale of renewable energy development foreseen under current UK energy policy.

The purpose of these two extreme models is to demonstrate that even if we opt for coal and/or nuclear this makes little difference to the energy equation over a period of eighty to one hundred years. All that developing a greater nuclear and coal capacity achieves, on the back of a declining oil and gas supply, is to make the peak in total energy availability to the UK economy a little higher, and the reduction in energy supply from 2030 to 2060 a little less steep.

In reality neither of these options is likely to be carried out. What will happen in the future is something that will be a synthesis between these two approaches. However, if we study the two extreme possibilities that face us today, as we begin to approach Peak Oil, we can take steps to ensure that the decisions we take in the near future do not store up problems for the next, and future, generations. Of course, we could "burn everything" and then follow the "reduce and renewable" strategy in a hundred years time in order to preserve a small proportion of our current energy supply. But this is not the point these models are seeking to illustrate. Under the "burn everything" strategy once the mineral resources are used up they have gone, forever, but if we were to follow the "reduce and renewable" strategy some use of these minerals (most likely the coal, with carbon sequestration to reduce carbon emissions) could be sustained for a few hundred years if we were able to significantly reduce consumption in the near future.

In reality we are not looking at an either/or decision. Over the next ten to twenty years we, as a nation, will be taking decisions on energy that could take us down one or other of these routes. What is important is that in advance of taking such decisions we are aware of the potential consequences twenty, or fifty or one hundred years into the future.

Either way, whether we burn it all or conserve as much as we can, by the beginning of the next century, the population of the globe might only have *one quarter* to *one-third* of the energy that we use today. Without knowing more about global energy use that might sound alarming. So, in terms of the UK population, what does one-quarter the level of our current energy consumption look like?

From the IEA's global dataset, the UK's current per-capita primary energy supply is 4toe (4 tonnes of oil equivalent – toe) per person [IEA, 2003]. But if we look at the figures for other countries, one-quarter of our current primary energy supply (1toe/person) is still quite a lot of energy. It's slightly more than the current level in China (0.9toe/person). It's a lot more than Georgia (0.46toe/person) or Kyrgyzstan (0.45toe/person), which climatically are more representative of the UK (hot Summers/ cold Winters). And it's way more than Bangladesh (0.15toe/person), although Bangladesh isn't really representative because they don't have such cold weather. And, although one-quarter of our current energy supply may seem small, 1toe per person per year is still more than the average for the whole of Asia (0.6toe/person) and Africa (0.63toe/person), and about the same as Latin America (1.07toe/person). So, it's not possible to say that reducing energy use by 75% is impossible, or would mean the end of our civilisation, because a large number of people around the world already live at or below this level of energy consumption.

The best way to understand how lower energy levels will affect our society is in terms of a slow progression over the space of two generations. We will change from a country that is materially rich to one that will be materially poorer (we will have less things) but which may retain the same diversity of learning and communication opportunities through the use of information and communications technology (ICT). This process of change will have to start within the next ten years, when the price of energy, and hence the goods made with materials and energy, starts to rise. Shortages, which will affect what goods are available, will begin to take hold from between 2025 and 2045 – depending upon which of the two extreme energy models we eventually follow. This means that most of the population alive today will see the decline of energy use begin to take hold. Severe shortages will grow during the second half of this century. So anyone under the age of twenty-five today is likely to see the new low-energy world take shape. But it is the children that are being born today who will, at the end of their lives, be living in this new low-energy world!

10. Tune Out, Turn Off, Be Happy?

If we take a mythical analogy, did Noah start building the Ark when the waters were rising? No, he had prior warning. But, given the difficulties of building the Ark and the fact that a flood big enough to consume the world mustn't have seemed that likely, why did he bother? This story, which some attribute to the flooding of the Black Sea basin over seven thousand years ago being recorded in the human consciousness, holds a message for us today. If the information that you are hearing sounds convincing enough, why wouldn't you act upon it, even though such a course of action might cause some short-term difficulty? Can we convince modern society that energy will begin to run short very soon? Or, like Noah's Ark, will the tribulations of the next one hundred years be recorded as another myth that human society, a few millennia from now, will use as an instructive moral tale?

In terms of the debate on Peak Oil, and what it means to society, it is interesting that in the UK there has been no meaningful political debate on the matter. We could also ask why, after forty years of public debate on the issue of the environment and human development, we are still very little further forward than the state of the world in the 1960s.

Today, the world uses a lot of fossil fuels. The UK reflects this high use of fossil fuels as part of its own energy economy. The scale of the UK's use of energy is not as high, in per-capita terms, as states such as the USA and some other states within Europe. The problem is that the current efforts to restructure energy use in order to meet climate change objectives does not necessarily reduce energy use – it just changes where we source the energy from. In fact the UK's use of energy is projected to increase at a steady rate over the next twenty years. This creates a problem for the UK. It needs to switch to other sources of energy to preserve its security of supply, and to avoid the cost implications of price rises following Peak Oil. But at the same time energy use is increasing, largely due to the unwillingness of government policy to intervene in the energy market [Guardian, 2004s].

What's worse, the current tokenism towards renewable sources of energy, especially small-scale renewable technologies not connected to energy grids, and the use of energy reduction policies, is storing up problems for the future. Renewable energy sources are being developed in the UK, and ostensibly this is part of a national policy strategy. The problem is, all that new renewable energy projects do is to meet a fraction of the projected increase in primary energy demand. They are not replacing energy from fossil fuel sources at all. This is, in part, due to the sceptical view taken of the role of renewable gy. A scepticism based not on the viability of the technology, but the fact that renewable sources cannot produce the same scale of energy production compared to fossil fuels.

If we take the "Reduce and Renewable Model" presented in the previous chapter, which is the only approach that would offer some form of energy supply in perpetuity, the results indicate that the UK must plan to use as little as one-quarter of the energy that is available to our society today. Reducing our economy to this level of energy use is not catastrophic in terms of other human societies. There are millions of people around the globe who already live with this level of energy consumption, or less. But even though it's not a catastrophe in terms of human society, it's going to spark a heated debate because Western society has grown used to material wealth – the product of three hundred years of development, colonial expropriation, and more recently the effective control of the global economic system. As a result there are many interest groups who have a driving mission to prevent any change that might upset the economic status quo amongst the industrialised nations.

To address these issues we need to change the nature of the debate about energy, and its relationship to the society we live in, within the UK political system. We must move beyond just counting kilowatts to look at the wider impacts of society having to accept a large-scale reduction in the use of energy over the course of the next century. The significant questions are "can it be done?", and "will people do it?". On past experience of how the world reacts to pressing environmental and social issues, the process is likely to be difficult to expedite because it will require fundamental change to the world's economic structures.

There have been similar warnings of resource crises in the past and, apparently, these portents of doom haven't materialised. However, in many ways this is self-fulfilling. As identified by Peter Chapman in the mid-70s [Chapman, 1975], claims about dire economic or environmental crises tend to be the victims of a paradox created by the way our society deals with sudden crises. If the problem is dismissed, then it's possible that the problem may arise as predicted. But in reality different sections of society do absorb the message, and by subtly changing the way society works the basis of the prediction is changed and the crisis is never realised.

However, it is an arguable that in the case of Peak Oil, or Peak Gas, the crisis is not avoidable. We have scoured pretty much the whole planet in search of natural resources, and we can be reasonable sure, as shown by the methods of M. King Hubbert, that a shortage of essential energy resources is imminent because of the scale of global energy use today. Despite the rights or wrongs of the warnings of dire catastrophe, the fact that various people have tried, and failed, to deal with resource- and development-based crises issues like Peak Oil can provide a valuable insight into how we might address this debate today.

The growth of environmentalism, from the 1960s, was based on a realisation that the economics of modern energy and resource use which underpin the consumer society are unsustainable. It began to enter the public consciousness with books like Rachel Carson's *Silent Spring* [Carson, 1962]. One of the leading figures on the debate over resource use during the 1960s was Paul Ehrlich. His book, *The Population Bomb* [Ehrlich, 1968], linked the growth of human society and resource use, and predicted dire consequences as the everyday materials that society relied upon ran out. This didn't happen, not because the theory was wrong, but for two other reasons. Partly, because a large section of the world population didn't get rich enough to consume lots of resources – the large-scale trickle-down of wealth from the rich to the poor nations forecast in the 1960s didn't materialise. But the main reason was that the data upon which Ehrlich's claims were based was incorrect. During the intervening years the development of new geophysical techniques allowed more land to be explored, and more resources to be found. At the same time new technology allowed greater efficiency savings.

However, with the greater level of data about natural systems we now possess today we can foresee that Ehrlich's thesis may come to pass over the next thirty to fifty years – primarily through depletion of energy resources by the expansion of Western consumer societies rather than population growth across the world as a whole. Technological improvements, which may slow these trends, are also reaching their physical limits, and science itself may be reaching the physical limits to the exploration of the natural world [Horgan, 1996].

We now possess, from the world's scientists, a prior warning of the imminent peak in oil production, and consequently, we can envision what may happen to modern society after that date. We also have prior warning of climate change and abrupt climate change, and what may happen if we don't stop burning fossil fuels. So, like Noah, our leaders are developing alternatives that seek to avoid the use of petroleum and solve climate change... but in reality, they're not. Just the opposite in fact.

Petroleum use is growing as countries like China and India industrialise. Energy use in the UK, driven by energy intensive activities such the growth in road and air transport, is also growing, and the UK government has no plan to curtail these trends. One reason for this is that our development of alternatives to the perceived crisis is being obstructed by political lobbyists. They believe that any change which affects the bottom line is bad for them, and due to their own importance, bad for society. So no-one should do anything that affects the level of economic activity until there's absolute proof of harm. The *p*roblem is, if you can't act upon knowledge-based precaution, and you must wait for certainty, how can you act until it is too late? As documented by observers of the environment movement, particularly in the USA where corporations have a higher profile in public debates [Klein, 2000], the inertia within government policy created by such vested interests is damaging the well-being of society.

In the early 1970s, the United Nations began its own studies into the effect of human development on the environment with the first Earth Summit [Ward, 1972]. At the same time, E.F. Schumacher, an economist from the UK's National Coal Board, promoted a whole new way of looking at human systems in his influential book *Small is Beautiful* [Schumacher, 1973]. The whole basis of modern society was under question by those sceptical about the value of technological society. Some of those who heard the message went as far as dropping out of modern society in search of alternative ways of living, for example, the self-sufficiency movement that developed around the work of John and Sally Seymour [Seymour, 1973].

However, this process was not one-way. By the 1980s and 1990s, as environmental legislation was beginning to affect the way that businesses operated, the perceived anti-development basis of environmental philosophy faced an onslaught from *wise use* lobbyists [Beder, 1997]. Wise use is a corporate-centred philosophy of resource and energy use, closely linked to the development of globalised markets. It advocates the un-restricted use of natural resources within a free market because it benefits society by creating wealth. Recent books such as *Small is Stupid* [Beckerman, 1995], *Hard Green* [Huber, 1999] and *The Skeptical Environmentalist* [Lomborg, 2001] sought to directly contradict the message of Ehrlich, Schumacher, and other advocates of more sustainable patterns of development.

The primary difference between the approach of wise use and environment groups is the data that they quote in support of their cause, or to refute the claims of their opponents. And although wise use lobbying may seem a rather futile activity in the face of public opinion on the environment, the fact that a one or two year delay in environmental regulation can save polluting industries millions makes it worthwhile. This means that for industrialists handing out a few hundred thousand pounds to a group of contrarian scientists to lobby on your behalf represents a good investment [Rowell, 1996].

The problem with those advocating change in the face of any type of environmental hazard, and which wise use groups exploit, is the *Great Aunt Principle*. For example, take smoking. Everyone seems to know someone – for example, a Great Aunt – who smoked forty-a-day and never had so much as a cold. And in fact, that's statistically true, and it can be shown as such for an individual case. But such an analysis can't be legitimately extended to society as a whole.

Genetic variability between humans means that while some people get sick the moment you spray a chemical-based perfume near their face, others can happily play with toxic chemicals till they're seventy with few ill effects. In the same way we have different faces or different hair colours, so genetics gives us a different susceptibility to environmental insults (air pollution, toxic chemicals, pollen, etc.). It's a statistical probability as to how many people within a given population will experience a harmful effect from some form of environmental insult, but you can't generalise on the basis that one person is unaffected by a particular hazard that the same result can be expected in all cases. In scientific terms, absence of evidence is not absence of effect whilst there is a strong case to show that the theory you are testing is correct.

The problem with the *Great Aunt Principle* is that it seeks to reduce the operation of statistical probability within complex systems to an absolute. It seeks to reduce knowledge to right or wrong, black or white answers when in reality there are a range of 'correct' answers based upon people's individual circumstances. Many people have a problem with scientific uncertainty, perhaps because knowing how uncertain something is seems contrary to the notion that scientific understanding provides certainty. It is for these reasons that those who wish to block a science-based argument for change can raise doubts over proof, even in the face of strong evidence to the contrary, as an argument for delaying action.

Those who wish to continue emitting greenhouse gases, or using toxic chemicals, or planting genetically-modified crops, use a more complex form of the *Great Aunt Principle*. In such an evaluation the Great Aunt is a section of the general public included in their study, and the 40-a-day cigarette habit is the questionable process which they seek to continue. And in the same way that genetic variability gives some people a strong immune system, there will undoubtedly be people who may benefit from the questionable activity, and who will provide the evidence that supports the claim. How representative this section of people are to society as a whole is always debatable [Stauber, 1995].

Wise use lobbyists will also argue that certain types of activity are essential to the well-being of society because without them society would be poorer. This is justified by allocating values to people, or plants, or animals, which are then factored into an economic case as to why certain forms of action are too expensive to contemplate. But this pre-supposes that these activities have always been carried out in the past, they always will be carried out in the future, that they have always benefited society in the past and will continue to do so in the future.

In the real-world the analysis is often more complex than the *Great Aunt Principle*, and it will be closely related to the motives of those involved. For example, if you have a fifty year-old chemical plant that's paid its capital costs decades ago every penny it makes above the operating cost is pure profit. This gives you an economic advantage. For this reason companies who operate with established plants, or in established industries, will resist change because it means new investment, new practices, and consequently they will have less return on their investment.

It's precisely this kind of political and economic inertia that is stalling the debate on reducing greenhouse gas emissions, or energy reduction. It's the same in relation to other international agreements on environmental or social conditions. For example the negotiations over climate change, or the reduction of polluting emissions, are stalled by arguments over certainty, uncertainty, or the health of Great Aunts. By contrast, the proceedings of the World Trade Organisation may be seen to operate at an almost dizzying pace as their development agenda supports the expansion of the market.

In part, the malaise over environmental and energy policy in the UK has been the responsibility of the environment movement itself. As the movement has become established, and networks of spare-bedroom activists have grown into multi-million pound organisations, the message that the original movement sought to communicate has been lost in the noise of the modern media society. At the same time, the way that statistics are used, or abused, by the media is obscuring the debate on environment and development issues. The mainstream environmental lobby groups have also lost the broad vision of how energy and resource use are the basis of nearly all environmental problems – from traffic growth to toxic pollution. Despite the improvements in knowledge and data collection since Ehrlich's original warnings, the way environmental issues are promoted by the mainstream environment organisations is masking the underlying argument – *consumerism is unsustainable.*

The problems of Peak Oil, and of climate change, are leading some to seriously advocate that the world faces an irresolvable crisis. This is partly because it's not just oil that is in short supply, but other resources too. These global problems, in particular access to water, land and other essential resources, have been outlined in detail by the United Nations Environment Programme in their *Global Environmental Outlook* reports [UNEP, 2002]. In particular, their Third Report outlines the vulnerabilities of human populations, particularly in developing countries, so that even minor changes in the quality of their environment may cause serious damage to their well-being. The international media are now highlighting the possibility that wars will soon begin over resource shortages – for example, the recent arguments on use of the waters of the River Nile over which Egypt said it would go to war with neighbouring states [EAP, 2004].

Some enlightened politicians are also beginning to raise concerns. In his book, *The 2030 Crunch* [Mason, 2003], the former Australian Senator Colin Mason argues that geopolitics (satisfying the land or resource needs of one state by removing the rights of another) will always take over from negotiated settlements when resource shortages begin to affect economic development. For this reason, he argues, we need stronger international institutions to manage the globe's resources. Whilst such globalist approaches are fine in principle, if we need a system that promotes more localisation how do we ensure equitable representation between such globalist controls and the more localised management of resources?

What we need to do is advance these debates past the negative concepts of *crunch* or *catastrophe*, and instead look to more positive ideas such as change and development. What we need is a *philosophy* – a set of common understandings of the physical and social phenomena that confront us, and which can look beyond the narrow perceptions of consumer-environmentalism, conservative-conservationists and the fatalism of the doom-mongers (by doom-monger, I refer to those who believe there is no solution, or the solution is business-as-usual). Such a philosophy was called for by the early environment movement but today, for the most part, it has been lost in the debate as to whether one particular technological solution or another will solve the pressing issues of pollution, energy shortages and climate change.

The purpose of a common philosophy is not to provide the answers to our questions, but instead to provide a framework where problems can be interpreted, and the questions examined and understood more fully. This allows us to move on towards solving the questions in the future in a way that addresses the root causes of the problem, not just the symptoms of the problem we see in the world today. More importantly a common philosophy would allow us to understand, in precisely the opposite way to those using the *Great Aunt Principle*, that there is no right or wrong answer to the problem. There will be a range of solutions depending upon local circumstances.

At the personal level such a philosophy allows the individual to question the basis of why things are as they are. Through this challenge to the existing perceptions of why these problems exist, they gather wisdom not about the answers, but about the reasons why these problems exist the first place. It is only by gathering such wisdom that the individual can gain an understanding of the issues, and through this understanding confidence in the reasons why we have to change. It is this step, seemingly lost on the solution-driven world of mass media politics, that the mainstream environment movement is failing to grasp. Without such confidence in the need for fundamental change, as society moves into what may seem in the short-term to be an insoluble crisis, the individual can have no belief that the solutions are feasible. And given the scale of change that confronts us, it is only by having confidence that the alternatives are feasible that people will be able to commit to move from a high to a low energy lifestyle.

So, let's define this philosophy. What's important is that is you can define a set of principles that must be part of a low energy society because there's no other way of doing things. Then, depending on circumstances, how people move from here to there is a matter of discussion and negotiation. What's important to understand is there is no one solution. If society seeks to live within the limits of the energy and resources provided by natural systems then a range of solutions must be applied because the natural environment varies from place to place, even in somewhere as small as the UK.

We can begin at first principles with a set of minimal standards that a lack of energy resources will enforce on society. As outlined during the early chapters of the book, you can't escape the laws of matter and thermodynamics. If you don't have the energy then you have to think of other ways of doing things – wishing will not make energy appear. So by assisting people to gain an understanding of what energy is, and how we as individuals relate to our use of energy, it is possible to explore how our existence depends upon the use of energy. And consequently, how that existence might or must change in order to carry it out at a lower level of energy intensity.

Different energy sources bring with them different costs and benefits. How we evaluate these costs and benefits is very subjective, and will largely depend on how each physical feature of a particular energy source is measured and evaluated. The greatest dis-benefit of the mineral energy resources – oil, coal, uranium, etc. – is that they are finite. Today our knowledge of the world is such that we can be reasonably sure that there are no large-scale undiscovered mineral resources, and that the resources we have left are not capable of meeting our ever increasing energy demands for more than a few decades.

World oil production is reliant upon low-cost primary sources of oil. Production of these deposits will peak over the next five to ten years and then decline steadily until they are exhausted. Following the peak in primary oil production – Peak Oil – secondary sources, like oil shales, tar sands and heavy oil, as well as tertiary sources like gas liquids and the production of liquid fuels from natural gas, will be far more important. However, these sources of oil cost a lot more to produce, and they will never be able to sustain the levels of production achievable from primary sources. With the increased costs of production, and with the changes to the volumes available to the world oil market, dealers will be bidding for the remaining productive capacity. Prices for oil are likely to rise two- or three-fold over the decade or so following the date of Peak Oil, which may in turn may create other economic effects. The rise in prices following Peak Oil could drive down demand as an economic recession, driven by price inflation, cuts in. Precisely how things will develop, when prices begin to rise, is not clear, and will be highly dependent upon the political situation at that time.

As an illustration of how fossil energy has insinuated its way into our lives, people should walk around their house, or their place of work, and identify those things which are *not* made with petroleum or hydrocarbon fuels (i.e., not made of plastics, glues and resins, chemicals, etc.) [BBC, 2002a]. Not made with, because today it's far harder to find things that are not influenced by the materials produced with fossil fuels than those that are. Even things such as wood or cotton, that are made from plants, are preserved, painted or dyed with chemicals made from natural gas or petroleum. Then, having identified just how much of your life is dominated by fossil energy, imagine what your life would be like if you doubled the cost of those products. What if petrol cost two pounds a litre instead of nearly one? What would be missing from your life if fossil-fuel-based goods rose twenty to fifty percent in price over the next ten to fifteen years? How much of what you have around you would you be willing to part with because it is unimportant to your life, and what would pay more for? Now double that price again (four times the current price) and make the same evaluation, as this is the sort of change you must envisage over the next two to three decades.

We know that oil and gas are running out and, from the research of M. King Hubbert and others in the petroleum industry, we have a good idea of when petroleum production will peak. So we can guess that in ten to twenty years time the goods you've identified as being valuable to your life, but which are highly dependent upon fossil energy, might cost two or more times their current price as the cost of energy rises. How much these price rises affect you depends upon what type of lifestyle you lead at the moment. If you are a self-sufficiency advocate, living on a smallholding in Wales, your main concern might be that the rise in taxes or the costs of the few items you must buy from the mainstream world might cost more than your frugal standard of living can afford. If you live in a small town you might be worried about paying for your energy bills and transport. If you live in a city, and your work and lifestyle are related to long-distance or international travel, eating out and the latest designer goods, then you are definitely in for a shock.

There are of course solutions to the oil, and then subsequently gas, running out, but as outlined earlier the solutions depend upon which development model we adopt. To maintain the current energy supply we could develop far more expensive energy production options. More precisely we won't develop them, as we already know how to build them because the technology has been around for a while. What will happen is that the rise in energy prices generally will price-in currently uneconomic energy sources. For example, we could build clean coal plants that sequester the carbon dioxide emissions produced by coal burning for disposal in old oil wells, or we could expand nuclear power. However the problems with these business-as-usual solutions, as explained earlier, is that they have a finite lifetime based on the availability of the mineral resources they are reliant upon. Other business-as-usual options, such as nuclear fusion or fast breeder reactors, although portrayed as reliable energy technologies are still largely theoretical. We can build the experimental plants, but getting such experimental designs to operate safely and produce a reasonable energy return on the energy invested in their operation is something that's yet to be perfected.

The alternate course from burning coal or uranium is to develop renewable energy sources. This would require that we cut our energy use by up to 75% over the next fifty years to roughly 1 tonne of oil equivalent (1toe, or 41,868MJ) per year because renewable energy sources are less dense energy sources than mineral energy sources. Producing our current level of energy consumption is not feasible because we would have to devote such large areas of land to collecting these less dense energy sources that it would preclude other activities essential to society – like growing food.

However, even though these energy sources are less dense they can be better suited to our needs. In our homes most of the energy we use is heat, even though it might be provided by other sources such as electricity or gas. Solar power (mostly thermal solar) would provide a ready source of heat energy in the Summer, and storing heat doesn't require expensive batteries or fuel cells just a large tank of water. In the Winter, provided that your home is well insulated, you might only use 10,000MJ to 15,000MJ of heat provided by other sources. That's low enough to be provided from a mixture of biomass-fired community CHP and improved heat from ground or water sourced heat pumps. What will need to be provided externally for most of the year, to supplement the energy balance of the house, is electrical power. This could be locally sourced from wind, biomass or biogas. The thing is when you cut energy consumption by 75% renewable energy options become far simpler, smaller scale, and easier to manage than when you are trying to run them in competition with dense mineral energy sources.

Living on 1toe of energy per year is not impossible. In fact, there are plenty of people in the world already living at this level of energy consumption or far less. The problem is that by our current standard of energy and resource consumption we would class these people as being "poor". It's at this point we may hit a snag. Our political classes have a great problem promising the people that they will be poorer in the future.

A good example of this is the UK's Sustainable Development Strategy [HMSO, 1999]. Entitled *A Better Quality of Life*, it outlines sustainable development in terms of the prudent use of natural resources and the maintenance of high and stable levels of economic growth. The strategy, written in the language of the business management culture, is a good example of how the concepts developed within the environment movement have been subverted to reflect the needs of economic globalisation. Obviously, in a world where energy is running short, either of these two aims is going to be a pretty tall order. In fact, I would challenge the UK government to state how, given the energy situation fifty years from now, UK citizens will have (in terms of the standards of the consumer society) a better quality of life and high levels of economic growth.

Returning to our concept of a philosophy, how can we have a realistic debate on being energy poor? The simple answer is to redefine the term *energy poor* to be meaningful in terms of energy use. In eighty or ninety years from now there will be few energy poor people in the world because there will be few highly intensive sources of energy that can be exploited to make people energy rich. In this sense everyone will be equally energy rich. Although this may sound tortuous, the issue of the amounts of energy will not be as relevant as the extent to which people can control and operate their own energy systems, and thus are independent of external factors that might affect their lives.

When most of the dense energy resources are depleted or rendered unusable by climate change, the forms of energy we will have left will be more closely related to those used by people living with less than 1toe per year – and who we would describe today as energy poor. 1toe per year is roughly the energy used by the average car in the UK each year, or half the amount of energy used to heat and power the average UK home for a year. But as that 1toe will have to cover everything, including agricultural and materials production, it means that we have to be very selective about the methods we use to get the resources we need to live. The other solution, to return to the studies by Paul Ehrlich, would be to forcibly reduce the population through birth control, as has been the case in China recently. However, this would be difficult to enforce for many reasons.

People who do live on 1toe or less of energy each year have something of great value that we don't possess – practical knowledge. This is something that many of those promoting renewable energy ignore. However, it is something that early environment movement guru John Seymour, and the Nineteenth Century author who Seymour based some of his work upon, William Cobbett [Cobbett, 1822], understood well. As Western society has become more technical, and we have become more knowledgeable about certain subjects, as a society we have become de-skilled about existence in general. We have lost the ability to manage and maintain the fabric of our lives without external help. So, how do we, as energy rich Westerners, learn to live like energy poor? The simple answer is that we need systems that allow people to access information to help them change their lifestyles through practical learning and experimentation, and get help with problems when they occur from informal sources (rather than paid expertise).

We also need systems that allow people to share knowledge, and through the sharing of their practical experience, to develop the level of collective knowledge further. In the past this would have required the use of expensive and energy intensive systems to support the printing and distribution of books and other literature. Today we have a system that is far more flexible, far more participative, and far less expensive in terms of energy and resource use – *information and communications technology*, or ICT. Telephones and computers do not move material, they move electrons – which are very light. They can also copy things over and over again using a minute fraction of the energy that would be use to make a paper-based publication, or plastic-based tape or CD containing video or sound.

A 1994 study in Japan showed that the average 61 hours of calls made from the average telephone each year used 123MJ of electrical power [Weizsäcker, 1998]. That's around the same energy that goes into the manufacture of 5 kilos of paper products, not including the transport, that the average UK household currently consumes in a two or three day period. Even if we scale-up the level of use to take account of the information we might send or receive over a year, electronic networks use a fraction of the energy that the printed media would use.

What telecommunications networks provide is a means whereby people can travel less, but they can still communicate and exchange information. This means that it will be possible to set up formal (distance learning) and informal (email and online gossip) networks where people can teach and support each other. So, rather than the knowledge and experience of those who successfully take-up low energy lifestyles being locked-up inside their heads, via electronic networks that knowledge can be shared for a fraction of the energy, and a fraction of the price, of using conventional means of communications such as books, the postal system, or printed newspapers. The problem with this approach is that computer equipment is highly energy intensive to make, and increasingly to use. This is largely due to the structure of the computer hardware and software industry at present. By endlessly making the programs people use demand more computer power, in the process introducing new features that render the older equipment obsolete, the industry forces computer users to buy new systems every two to three years.

The built-in obsolescence of the computer industry, and other consumer-based technological systems, is not an inviolable rule. There are currently many projects around the world, some official and some ad hoc, seeking to develop an alternative to this highly resource intensive model of computer use. One of the most notable projects to date, created as global collaboration from within the Internet community, is the Gnu/Linux computer operating system. This doesn't just provides an alternative to the industry-centred software systems that dominate the computer industry. It provides an excellent model for how we might develop complex technological systems in the future without the large scale, centralised investment of energy and resources that has been the model for technological progress since the Industrial Revolution.

In turn, the Gnu/Linux system has helped spawn many more projects and support networks that are encouraging people to refurbish and use older computer equipment productively. The primary benefit of this, particularly to developing countries, is that it avoids the high costs of purchasing new equipment [Free Range, 2004]. Following this open model, pioneered by the Gnu/Linux operating system, the costs of computer equipment and software would fall significantly. At the same time, the emphasis on providing affordable equipment with a long service life would cut the overall energy intensity of using information and communications technology.

The other thing that electronic networks can help organise at the local level is social activity, and especially local trading networks. In a system where work, energy and food production is locally sourced, it might be that there's little need to use money as the basis for exchange – something that is is common in those parts of the world that use less than 1 toe per year. Instead people trade goods or labour. Local networks enabled by online systems could support this type of low-cost trading by providing an easily accessible directory of what people need, and what people have to offer. The obvious problem this creates is, where does the tax come from? Even a low energy society would still need to organise some services on a regional or national basis. The solution, rather than taxing income, would be a tax on resource use – such as paper, steel or certain types of fuel – used in the production of new goods. In this way everyone pays tax according to the level of new resources they consume. This would also encourage a high level of reuse and repair to avoid paying the higher replacement costs of virgin material.

One other block to getting from here to there are the economic and legal principles that have been developed during the Industrial Age – such as global markets, intellectual property rights, and centralised economies. These would have to be redefined, changing the balance of the law towards local-level organisations. By decentralising the basis of economic organisation it makes it possible for energy and resources to be sourced locally, using renewable energy and locally manufactured goods. Likewise intellectual property rights, developed largely as a result of the economic controls sought over the industrial society, stifle the ability of people to modify and adapt ideas, and to copy information that might help and support the work of others. This is especially true of a society where the main means of moving information will be computer networks, and where controls over the transmission and use of information would hinder our development towards a low energy society. The restrictive use of intellectual property rights, and many of the other legal restrictions that would hinder greater localisation and local development, will need to reviewed.

A low energy lifestyle, powered by renewable energy and some manufactured energy (like biofuels), is entirely possible using existing technologies. It is in no way experimental. What would be experimental, in forty to sixty years, is getting an island of sixty to seventy million people (the UK) to do this at the same time. The problem is that, from the point of view of people standing in the middle of a large urban area today, this may seem impossible. And it is this mental gap, between people's understanding of the world today and their confidence that we can change the structures of our society over the next forty to fifty years, that we have to bridge in the near future in order to give people the confidence to change the way society operates.

The great problem in advocating a solution to Peak Oil, and the perceptions of an energy crisis that will follow shortly behind, is that there is apparently no solution within our current conception of the high-energy modern society. This is the wrong approach. In reality there is no one solution, but a diversity of options. There will be a range of options depending upon how people would like to live and work in a future low-energy society. For some, that may by the kind of low-energy, back-to-the-land type of lifestyle envisaged by the self-sufficiency movement of the 1970s. For others it will mean living in low-energy urban communities where most of the technical or energy intensive manufacturing work will be based. For most, it may be somewhere between these two states. How people will chose to live will largely depend on the process that takes them from the high-energy state to the low-energy state over the next fifty to eighty years. It's also highly contingent upon the effects of international tension over resources, and the effects of climate change, and especially the growth in the movement of environmental refugees, that that is likely to develop over this period.

Only two things are certain in this process:

1. The amount of energy available to society will start to run down, soon, and whether we take the reduce and renewable or burn everything option we end up in roughly the same energy situation in one hundred to one hundred and twenty years time. So which option you chose – reduce and renewable or burn

everything – makes little difference in the long-term, but it does make a significant difference in terms of the confidence or security that could be designed into a more managed transition in the short-term.

2.   Following the peak in conventional oil production, sometime in the next five to ten years, world energy prices will begin to rise and they will not fall again. Which, in an global economy built upon the principle of cheap energy, is going to be as disruptive in the short-term as energy running out will be in the long term.

Given what we know about the oil economy today we shouldn't be worried about the oil running out, but the fact that it will get very expensive long before then. This in turn will make the globalised economy of the industrialised nations unstable because these nations depend upon a cheap and plentiful supply of energy for their continued growth. For those people who don't wish to directly participate in this system there is only one possible means of action. You have to think of obtaining your energy "beyond oil".

The purpose of this book has been to explore the linkage between the way we live our lives today and the use of dense energy sources. It proposes no detailed solutions, because there is no detailed solution that will work for everyone. Therefore we have to take a greater interest in, and responsibility for, how we might deal with the problems that develop over the next twenty to forty years as the world's energy supply contracts.

Anyone who wants to plan how they might manage their own transition to a low energy lifestyle has a simple decision to make almost immediately – *where and how do I want to be living in ten years time.* This is because what you do today, in terms of investing in energy efficient energy systems, or attending courses to learn how to develop your own small-scale energy systems, has far more relevance to how you might deal with the contraction of our energy supply in the future. Unless we, as individuals and as a society, take the time to understand and plan our transition to a low-energy lifestyle today, following Peak Oil, and more particularly following Peak Gas, we are going to have a far more abrupt revelation to the value of energy within our lives.

Sources and Further Information

The references used in this book have been selected not just because of the information they contain, but also because they are for the most part freely available. For example, the *BBC* and *The Guardian* are quoted extensively as sources of up to date news. This doesn't necessarily reflect the quality of their output in general, but the fact that they run excellent news web sites that provide information free of charge, and without the troublesome registration systems that result in your personal data being sold around the globe.

To make life easier all the sources listed below are available online via the book's web site. This not only means that to go to a source you just click a link (no tedious typing required), but should the location of the source change the web page can provide the most up-to-date location of the information, or an alternative source that provides similar information. For certain books, if they are in print, links are also given to online bookshops where you can buy a copy.

Following the book's sources, listed below, you will also find more general information on where you can obtain more up-to-date information on energy issues. This information is also provided via the book's web site:

http://www.fraw.org.uk/ebo/

Note that the links given for the references below were correct in January 2005.

BBC, 2002a   Science at Nine – *Crude Facts*, BBC Radio 4, 19th July 2002. http://www.bbc.co.uk/radio4/science/crudefacts.shtml

BBC, 2002b   The Westminster Hour – *Dennis Sewell on the Think Tanks*, BBC Radio 4, November/December 2002. http://news.bbc.co.uk/1/hi/programmes/the_westminster_hour/2551927.stm

BBC, 2003a   *2003 Climate Havoc Cost $60bn*, BBC News, 11th December 2003. http://news.bbc.co.uk/2/hi/americas/3308959.stm

BBC, 2003b   Horizon – *The Big Chill*, BBC2 TV, 13th November 2003. http://www.bbc.co.uk/science/horizon/2003/bigchill.shtml

BBC, 2003c   The Money Programme – *Blackout Britain*, BBC 2, 12th October 2003.

BBC, 2003d   Costing the Earth – *The Hydrogen Bubble*, BBC Radio 4, 18th December 2003. http://www.bbc.co.uk/radio4/science/

costingtheearth_20031218.shtml

BBC, 2003e — The Food Programme – *Self Sufficiency*, BBC Radio 4, 18th May 2003. http://www.bbc.co.uk/radio4/factual/foodprogramme_ 20030518.shtml

BBC, 2004a — *Stability Fear as Oil Reliance Grows*, BBC News Online, Tuesday 26th October 2004. http://news.bbc.co.uk/1/hi/business/ 3953907.stm

BBC, 2004b — *World Oil Demand Estimate Raised*, BBC News Online, Wednesday 11th August 2004. http://news.bbc.co.uk/1/hi/business/3554462.stm

BBC, 2004c — *US Oil Output Lowest Since 1950*, BBC News Online, Tuesday 9th November 2004. http://news.bbc.co.uk/1/hi/business/3997589.stm

BBC, 2004d — *Energy: Meting the Soaring Demand*, Alex Kirby, BBC News Online, Tuesday 9th November 2004. http://news.bbc.co.uk/1/hi/sci/tech/3995135.stm

BBC, 2004e — *UK Oil Balance Moves Into Deficit*, BBC News Online, Tuesday 9th November 2004. http://news.bbc.co.uk/1/hi/business/3995719.stm

BBC, 2004f — *Climate Wars*, BBC Radio 4, 13th January 2004. http://www.bbc.co.uk/radio4/science/climatewars.shtml

BBC, 2004g — *Scorching World of Climate Politics*, BBC News, 12th January 2004. http://news.bbc.co.uk/2/low/science/nature/3390901.stm

BBC, 2004h — *Micropower 'Could Fuel UK Homes'*, BBC News Online, Tuesday 14th September 2004. http://news.bbc.co.uk/2/hi/ science/nature/3650208.stm

BBC, 2004i — *If... the lights go out*, BBC 2, 10th March 2004. http://news.bbc.co.uk/1/hi/programmes/if/3487048.stm

BBC, 2004j — Costing the Earth – *Building a Greener Home*, BBC Radio 4, Thursday 11th April 2002. http://www.bbc.co.uk/radio4/ science/costingtheearth_20020411.shtml

BBC, 2004k — *Power cuts could hit UK by 2006*, BBC News, 10th March 2004. http://news.bbc.co.uk/1/hi/business/3496844.stm

BBC, 2004l — The Food Programme – *Ready Meals*, BBC Radio 4, 8th February 2004. http://www.bbc.co.uk/radio4/factual/ foodprogramme_ 20040208.shtml

Beder, 1997 — *Global Spin – The Corporate Assault on Environmentalism*, Sharon Beder, Green Books, 1997. http://www.uow.edu.au/arts/sts/ sbeder/global.html

BedZed, 2004 — BedZed housing development brochure, The Peabody Trust, 2004.

http://www.bedzed.org.uk/

Biogen, 2002   *Anaerobic Digestion of Farm and Food Processing Residues: Good Practice Guidelines*, British Biogen, 2002.
http://www.britishbiogen.co.uk/gpg/adgpg/adgpgfront.htm

BP, 2004   *Energy In Focus – The BP Statistical Review of World Energy 2004*, British Petroleum, 2004. http://www.bp.com/subsection.do?
categoryId=95&contentId=2006480

Beckerman, 1995   *Small is Stupid – Blowing the Whistle on the Greens*, Wilfred Beckerman, Gerald Duckworth and Co. 1995.

Byron, 1816   *Darkness*, Alfred Lord Byron, first published 1816. http://quotations.about.com/cs/poemlyrics/a/Darkness.htm

Caltech. 2003a   *Hydrogen Economy Might Impact Earth's Atmosphere, Study Shows*, California Institute of Technology Media Release, 12th June 2003.
http://pr.caltech.edu/media/Press_Releases/
PR12405.html

Caltech, 2003b   *Potential Environmental Impact of a Hydrogen Economy on the Stratosphere*, Tracey K. Tromp, Run-Lie Shia, Mark Allen, John M. Eiler, Y. L. Yung, pp.1740-1742, Science, volume 300, no.5626, 13th June 2003. http://www.waterstof.org/
20030613POLEMIEK_ Hydrogen-Science-Article-June-2003.pdf

Campbell, 1998   *The End of Cheap Oil*, Colin J. Campbell and Jean H. Laherrère, pages 78 to 83, Scientific American, March 1998.
http://dieoff.org/page140.htm

Campbell, 1999   *The Imminent Peak of World Oil Production*, Presentation to a House of Commons All-Party Committee, 7th July 1999, Dr. Colin Campbell. http://www.hubbertpeak.com/campbell/commons.htm

Campbell, 2000   *The Myth of Spare Capacity: Setting the Stage for Another Oil Shock*, Oil and Gas Journal, 20th March 2000.
http://www.hubbertpeak.com/campbell/mythcap.htm

Carson, 1962   *Silent Spring*, Rachel Carson, 1962 (first published in the UK, 1963).
http://www.rachelcarson.org/

Channel 4, 2004   *Oil Supplies 'Over-Estimated'*, Channel 4 News, 26th October 2004.
http://www.channel4.com/news/2004/10/week_5/26_oil.html

Chapman, 1975   *Fuel's Paradise – Energy Options for Britain*, Peter Chapman, Penguin Books, 1975.

Climate, 2003   *Proxy climatic and environmental changes of the past 1000 years*, Willie Soon and Sallie Baliunas, Climate Research vol. 23, pp89-110, 2003. http://www.int-res.com/abstracts/cr/v23/n2/p89-

110.html

Cobbett, 1822	*Cottage Economy*, William Cobbett, 1822 – recently republished by Verey and Von Kanitz, 2000.
CorpWatch, 2000	*Carbon Capitalism*, CorporateWatch Magazine 11, 2000. http://www.fraw.org.uk/ebo/mirror/cw1cc1.html
CRE, 2001	*Energy Costs and Agriculture*, US Congressional Research Service, 24th April 2001. http://www.ncseonline.org/NLE/CRS/abstract.cfm?NLEid=23512
DEFRA, 2003	*Agriculture in the United Kingdom 2002*, Department of the Environment, Food and Rural Affairs, January 2003. http://statistics.defra.gov.uk/esg/publications/auk/2002/default.asp
DEFRA, 2004	*Consultation on the Review of the UK Climate Change Programme*, Department of the Environment, Food and Rural Affairs, 8th December 2004. http://www.defra.gov.uk/corporate/consult/ukccp-review/index.htm
DETR, 1999	*Transport of Goods by Road 1998*, Department of the Environment, Transport and the Regions 1999.
DfT, 2002	*Powering Future Vehicles Strategy*, Department for Transport, July 2002. http://www.dft.gov.uk/stellent/groups/dft_roads/documents/pdf/dft_roads_pdf_506885.pdf
DfT, 2003	*Powering Future Vehicles Strategy – First Annual Report*, Department for Transport, October 2003. http://www.dft.gov.uk/stellent/groups/dft_roads/documents/page/dft_roads_024731.pdf
Dickson, 2002	*Rapid Freshening of the Deep North Atlantic Ocean Over the Past Four Decades*, B. Dickson, I. Yashayaev, J. Meincke, B. Turrell, S. Dye, and J. Hoffort, Nature, vol. 416, 25th April 2002. http://asof.npolar.no/library/pdf/dicksonetal.pdf
DJNW, 2003	*DJ EU Cost Gearing Up for Emissions Trading: CO2 Price Rising*, Dow Jones Newswire, 19th November 2003. http://framehosting.dowjonesnews.com/sample/samplestory.asp?StoryID=20031119102 80003&Take=1
DoE, 1992	*Waste Management Paper No.1: A Review of Options* (second edition), Department of the Environment 1992.
Douthwaite, 2003	*Before the Wells Run Dry – Ireland's Transition to Renewable Energy*, edited by Richard Douthwaite, Lilliput Press 2003 (distributed in the UK by Green Books). http://www.feasta.org/documents/wells/contents.html
DTI, 2002a	*UK Energy Flowchart*, Department of Trade and Industry, July 2002.

http://www.dti.gov.uk/energy/inform/energyflowchart.pdf

DTI, 2002b — *Energy Consumption in the United Kingdom*, Department of Trade and Industry/National Statistics, 2002. http://www.dti.gov.uk/energy/inform/energy_consumption/

DTI, 2002c — *Energy Paper 68*, Department of Trade and Industry, 2002. http://www.dti.gov.uk/energy/inform/energy_projections/ep68_final.pdf

DTI, 2003a — *UK Energy in Brief July 2003*, Department of Trade and Industry and National Statistics, July 2003. http://www.fraw.org.uk/ebo/mirror/energyinbrief2003.pdf

DTI, 2003b — *Our Energy Future: Creating a Low Carbon Economy – Supplementary Annexes*, Department of Trade and Industry/ Department of the Environment, Food and Rural Affairs, 2003. http://www.dti.gov.uk/energy/whitepaper/annexes.pdf

DTI, 2004a — Table 5.1, *Commodity Balances – Electricity*, Digest of UK Energy Statistics (DUKES), Department of Trade and Industry, 2004. http://www.dti.gov.uk/energy/inform/energy_stats/electricity/dukes5_1.xls See also http://www.dti.gov.uk/energy/inform/dukes/dukes2004/index.shtml for detailed reports

DTI, 2004b — Tables 1.1–1.3, *Aggregate Energy Balances 2003*, Digest of UK Energy Statistics (DUKES), Department of Trade and Industry, 2004. http://www.dti.gov.uk/energy/inform/energy_stats/total_energy/dukes1_1-1_3.xls See also http://www.dti.gov.uk/energy/inform/dukes/dukes2004/index.shtml for detailed reports

DTI, 2004c — Table 1.1.1, *Inland Consumption of Primary Fuels and Equivalents for Energy Use 1970-2003*, Digest of UK Energy Statistics (DUKES), Department of Trade and Industry, 2004. http://www.dti.gov.uk/energy/inform/energy_stats/total_energy/dukes1_1_1.xls See also http://www.dti.gov.uk/energy/inform/dukes/dukes2004/index.shtml for detailed reports

DTI, 2004d — *Energy – Its Impact on the Environment and Society – 2004 update*, Department of Trade and Industry 2004. http://www.dti.gov.uk/energy/environment/energy_impact/index.shtml

DTI, 2004e — Table 7.1-7.3, *Commodity Balances 2003 – Renewables and Waste*, Digest of UK Energy Statistics (DUKES), Department of Trade and Industry, 2004. http://www.dti.gov.uk/energy/inform/energy_stats/renewables/dukes7_1-7_3.xls

See also http://www.dti.gov.uk/energy/inform/dukes/dukes2004/

index.shtml for detailed reports

DTI, 2004f   *Renewable Innovations Review*, Department of Trade and Industry, February 2004. http://www.dti.gov.uk/energy/renewables/policy/ renewables_innovation_review.shtml

DTI, 2004g   *Renewable Supply Chain Gap Analysis – Summary Report*, Department of Trade and Industry, January 2004. http://www.dti.gov.uk/energy/renewables/publications/pdfs/ renewgapreport.pdf

DTI, 2004h   *UK Energy in Brief July 2004*, Department of Trade and Industry and National Statistics, July 2004. http://www.dti.gov.uk/ energy/inform/energy_in_brief/index.shtml

DTI-EST, 2002   *Get Solar PV – Your Electricity, Your Environment*, Department of Trade and Industry/The Energy Saving Trust, 2002. http://www.solarpv-grants.co.uk/ and http://www.fraw.org.uk/ ebo/mirror/solar_pv_flyer.pdf

DTI-EST, 2003   *The Major Photovoltaic Demonstration Programme*, Energy Saving Trust/Department of Trade and Industry 2003. http://www.solarpv-grants.co.uk/ and http://www.fraw.org.uk/ ebo/mirror/Fact-St1OffG_fin.pdf

E4Tech, 2003   *Biomass for Heat and Power in the UK – A Techno-Economic Assessment of the Long Term Potential* (Final Report), E4Tech for the Department of Trade and Industry, December 2003. http://www.dti.gov.uk/energy/renewables/policy/ e4techbiomass.pdf

EAP, 2004   *Water Wars Loom Along the Nile*, East Africa Post, 16th January 2004. http://www.news24.com/News24/Africa/News/0,,2-11-1447_1470431,00.html

Earth Cymru, 2004   *What a Gas! – how the development of liquefied natural gas facilities creates a more expensive and insecure energy future*, Earth Cymru Network Bulletin 3/04, October 2004. http://www.fraw.org.uk/ ecn/ecnb/ecnb-03_04.html

EC, 2000   *Towards a European Strategy for the Security of Energy Supply – Technical Document*, European Commission, 2000. http://europa.eu.int/comm/energy_transport/doc-technique/ doctechlv-en.pdf

EC, 2001   Energy Green Paper, *Towards a European Strategy for the Security of Energy Supply*, European Commission 2001. http://europa.eu.int/ comm/energy_transport/doc-principal/pubfinal_en.pdf

EC, 2002   *Energy – Let's Overcome Our Dependence*, European Commission

2002. http://europa.eu.int/comm/energy_transport/
livrevert/brochure/dep_en.pdf

Economist, 2001   *A Dangerous Addiction*, The Economist, volume 361 no.8252, 15th December 2001. http://www.economist.com/printedition/ displayStory.cfm?Story_ID=904425

EcoSec, 2002   *Dow Jones Emission Trading Scheme Could Be Used to Finance Asian Power Projects*, 6th February 2002. http://www.ecosecurities.com/ 200about_us/222world_press/222dj_6_feb_ 2002.html

ECSSR, 2003   *The Future of Oil as a Source of Energy*, The Emirates Centre for Strategic Studies and Research, 2003.

EEA, 2004   *The Impacts of Europe's Changing Climate – An Indicator-Based Assessment*, European Environment Agency Report 2/2004, 2004. http://reports.eea.eu.int/climate_report_2_2004/en/tab_ content_RLR

EHDRP, 2003   The European Hot Dry Rocks Project – http://www.soultz.net/

Ehrlich, 1968   *The Population Bomb*, Paul Ehrlich, Ballantine Books 1968.

ETSU, 1990   *Gaseous Emissions due to Electricity Fuel Cycles in the United Kingdom* (Energy and Environment Paper No.1), N.J. Eyre, Energy Technology Support Unit (ETSU) for the Department of Energy, 1990.

ETSU, 2001   *Potential Cost Reductions in PV Systems*, Arthur D. Little, Cambridge Consultants Limited, under contract to the Energy Technology Support Unit (ETSU) and the Department of Trade and Industry, 2001. http://www.dti.gov.uk/energy/renewables/ publications/pdfs/sp200320.pdf

Farmatic, 2002   *Anaerobic Digestion*, presentation to the Bath and West Winter Show, 21st November 2002, Farmatic Biotech Energy UK, 2002. http://www.bathandwest.co.uk/files/journal/AnaerobicDigestion/ AnaerobicDigestion.pdf

Free Range, 2004   *The Free Range Community-Linux Training Centre Project* – http://www.fraw.org.uk/cltc/index.shtml   and the *The Free Range Salvage Server Project* – http://www.fraw.org.uk/ssp/index.shtml

GBN, 2003   *An Abrupt Climate Change Scenario and its Implications for US National Security*, Peter Schwartz and Doug Randall of the Global Business Network on behalf of the US Department of Defense, October 2003. http://www.gbn.com/ArticleDisplayServlet. srv?aid=26231

Guardian, 2003   *Farmers Burned as Green Energy Plant Faces Export*, The Guardian, 31st May 2003. http://politics.guardian.co.uk/green/story

/0,9061,967595,00.html

Guardian 2004a — *An Answer In Somerset – The Age of Entropy is Here, We Should All Now Be Leaning to Live Without Oil*, George Monbiot, The Guardian, Tuesday August 24th 2004.
http://www.guardian.co.uk/oil/story/0,11319,1289537,00.html

Guardian 2004b — *Hedge Funds to Come Under Closer Scrutiny*, Larry Elliot, The Guardian, Monday 4th October 2004.
http://www.guardian.co.uk/oil/story/0,11319,1319111,00.html

Guardian, 2004c — *Britain's Bounty is Running Dry* (first of a 3 part series on UK's falling oil reserves), Ashley Seager, The Guardian, Tuesday 12th October 2004. http://www.guardian.co.uk/business/story/0,3604,1324879,00.html

Guardian, 2004d — *IMF Says Expensive Energy is Here to Stay*, Larry Elliot, The Guardian, Thursday 30th September 2004.
http://www.guardian.co.uk/oil/story/0,11319,1315971,00.html

Guardian, 2004e — *High Oil Prices Threaten to Stoke Inflation*, Ashley Seager, The Guardian, Tuesday 9th November 2004.
http://www.guardian.co.uk/business/story/0,3604,1346536,00.html

Guardian, 2004f — *Hedging Bets*, Terry Macallister, The Guardian, Wednesday 29th September 2004. http://www.guardian.co.uk/oil/story/0,11319,1315128,00.html

Guardian, 2004g — *G7 Backs Brown Plan to Ease the Oil Crisis – Reforms Urged as IMF Talks of High Prices Until 2010*, Larry Elliot, The Guardian, Saturday 2nd October 2004. http://www.guardian.co.uk/oil/story/0,11319,1318063,00.html

Guardian 2004h — *Climate Fear as Carbon Levels Soar*, Paul Brown, The Guardian, Monday 11th October 2004. http://www.guardian.co.uk/climate-change/story/0,12374,1324379,00.html

Guardian, 2004i — *Scientist Claim New Evidence of Warming*, Matthew Taylor, The Guardian, Thursday 6th May 2004. http://www.guardian.co.uk/international/story/0,3604,1210617,00.html

Guardian, 2004j — *Worries of Rising Carbon Dioxide Emissions*, Mark Milner, Wednesday 27th October 2004. http://www.guardian.co.uk/oil/story/0,11319,1336669,00.html

Guardian, 2004k — *First Things First*, Bjørn Lomborg, The Guardian, Thursday October 28th 2004. http://www.guardian.co.uk/life/opinion/story/0,12981,1337209,00.html

Guardian, 2004l — *This is Neither Scepticism Nor Science – Just Nonsense*, Tom Burke, Saturday 24th October 2004. http://www.guardian.co.uk/

climatechange/story/0,12374,1334274,00.html

Guardian, 2004m   *An Unnatural Disaster*, The Guardian, 8th January 2004. http://www.guardian.co.uk/climatechange/story/0,12374,1118281,00.html

Guardian, 2004n   *Blair Reignites Nuclear Debate*, The Guardian, 7th July 2004. http://www.guardian.co.uk/nuclear/article/0,2763,1255690,00.html

Guardian, 2004o   *Britain Missing Emissions Target*, Adam Jay, The Guardian, Wednesday December 8th 2004. http://politics.guardian.co.uk/green/story/0,9061,1369158,00.html

Guardian, 2004p   *The Age of Cheap Oil is Over*, Larry Elliot, The Guardian, Wednesday 1st December 2004. http://www.guardian.co.uk/oil/story/0,11319,1363296,00.html

Guardian, 2004q   *Hydrogen Seen as Car Fuel of the Future*, Paul Brown, The Guardian, Friday 10th September 2004. http://www.guardian.co.uk/renewable/Story/0,2763,1301402,00.html

Guardian, 2004r   Solar Power Sucked into Funding 'Black Hole', Mark Tran, The Guardian, Thursday 2nd September 2004. http://www.guardian.co.uk/renewable/Story/0,2763,1296018,00.html

Guardian, 2004s   *The Road to Nowhere*, John Vidal, The Guardian, Friday September 17th 2004. http://www.guardian.co.uk/waste/story/0,12188,1306507,00.html

Hallam, 2003   *Evaluation of the Comparative Energy, Global Warming, Socio-Economic Costs and Benefits of Biodiesel*, School of Environment and Development, Sheffield Hallam University under contract from the Department of Environment, Food and Rural Affairs, January 2003. http://www.shu.ac.uk/rru/projects/biodiesel_evaluation.html

Heinberg, 2003   *The Party's Over: Oil, War and the Fate of Industrial Societies*, Richard Heinberg, New Society Publishers 2003.

HMSO, 1994   *Climate Change: The UK Programme* (Cm2427), HMSO 1994.

HMSO, 1999   *A Better Quality of Life – A Strategy for Sustainable Development for the United Kingdom* (Cm4345), HMSO, May 1999. http:// www.sustainable-development.gov.uk/uk_strategy/content.htm

HMSO, 2000   Part 2, *Waste Strategy 2000* (Cm4693), HMSO, May 2000. http://www.defra.gov.uk/environment/waste/strategy/cm4693/index.htm

HMSO, 2003   *Our Energy Future: Creating a Low Carbon Economy* (Cm5761), HMSO 2003. http://www.dti.gov.uk/energy/whitepaper/index.shtml

Horgan, 1996	*The End of Science: Facing the Limits of Knowledge in the Twilight of the Scientific Age*, John Horgan, Addison Wesley, 1996 (published in the UK by Abacus).
Hubbert, 1956	*Nuclear Energy and Fossil Fuels*, M. King Hubbert, published in Proceedings of the Spring Meeting of the American Petroleum Institute, 1956.
Huber, 1999	*Hard Green – Saving the Environment From the Environmentalists*, Peter Huber, Basic Books 1999. http://www.hardgreen.com/
IEA, 2001	*World Energy Outlook 2001 – 2001 Insights*, International Energy Agency (IEA) and the Organisation for Economic Co-operation and Development (OECD) 2001. http://library.iea.org/dbtw-wpd/Textbase/npold/npold_pdf/weo2001.pdf
IEA, 2003	*Key World Energy Statistics 2003*, International Energy Agency (IEA) 2003. http://www.iea.org/statist/key2003/keyworld2003.pdf
ILEX, 2003	*Implications of the EU ETS for the Power Sector*, ILEX Energy Consulting for the Department of Trade and Industry and the Department of the Environment, Food and Rural Affairs, and OFGEM, September 2003. http://www.dti.gov.uk/energy/sepn/ilex_report.pdf
Jesch, 1981	*Solar Energy Today*, Leslie F. Jesch, UK Section – International Solar Energy Society, 1981.
JESS, 2003	*Second Report of the Joint Energy Security of Supply Working Group*, Department of Trade and Industry/OFGEM, February 2003. http://www.dti.gov.uk/energy/domestic_markets/security_of_supply/jessreport2.pdf
Klein, 2000	*No Logo*, Naomi Klein, Flamingo, 2000. http://www.nologo.org/
Laherrère, 2001	*Forecasting Future Production from Past Discoveries*, Jean H. Laherrère, OPEC Seminar, 28th September 2001. http://www.hubbertpeak.com/laherrere/opec2001.pdf
Laherrère, 2003	*Oil and Natural Gas Resource Assessment: Production Growth Cycle Models*, Jean H. Laherrère, draft copy from the Association for the Study of Peak Oil, 16th July 2003. http://www.hubbertpeak.com/laherrere/EncyclopediaOfEnergy.doc
LEK, 2004	*Review of the Economic Case for Energy Crops in the UK*, LEK Consulting for the Department of Trade and Industry, January 2004. http://www.dti.gov.uk/energy/renewables/policy/lekreview.pdf
Lomborg, 2001	*The Skeptical Environmentalist: Measuring the Real State of the World*, Bjørn Lomborg, Cambridge University Press, 2001.

	http://www.lomborg.com/books.htm
Lomborg, 2004	*The Copenhagen Consensus*, Bjørn Lomborg (editor), Cambridge University Press, 2004. http://www.copenhagenconsensus.com/
Lovelock, 1979	*Gaia – A New Look at Life on Earth*, James Lovelock, Oxford University Press, 1979.
Marshall, 2004	*Climate Change*, environment policy website maintained by the Marshall Institute. http://www.marshall.org/subcategory.php?id=9
Mason, 2003	*The 2030 Spike: Countdown to Global Catastrophe*, Colin Mason, Earthscan Books 2003. http://www.2030spike.com/
McKillop, 2003	*Price Signals and Global Energy Transition*, Andrew McKillop, International Association of Energy Economists, 2003. http://www.hubbertpeak.com/mcKillop/PriceSignals.pdf
Meyer, 2003	*Contraction and Convergence – The Global Solution to Climate Change*, Aubrey Meyer, Green Books, 2003. For detailed information on *Contraction and Convergence* see the Global Commons Institute website, http://www.gci.org.uk/contconv/cc.html
Mobbs 2004a	*Turning the World Upside Down*, p16, The World Today (Journal of the Royal Institute for International Affairs), vol. 60 no.12, December 2004. http://www.fraw.org.uk/ebo/papers/the_world_today.html
Mobbs 2004b	*An Analysis of Turbine Size and Power Production*, Paul Mobbs, Mobbs' Environmental Investigations, December 2004. http://www.fraw.org.uk/ebo/papers/wind_farm_analysis.html
NAO, 1999	*Audit of the Future Oil Price Convention for the November 1999 Pre-Budget Report*, National Audit Office, November 1999. http://www.nao.gov.uk/publications/nao_reports/9899873.pdf
NatGeo, 2004	*The End of Cheap Oil*, National Geographic, vol. 205 no.6, June 2004. http://magma.nationalgeographic.com/ngm/0406/feature5/
NERA, 2002	*Security in Gas and Electricity Markets – Final Reports for the Department of Trade and Industry*, New Economics Research Associates on behalf of the Department of Trade and Industry, October 2002. http://www.dti.gov.uk/energy/whitepaper/dti_security.pdf
New Scientist, 2003	*Greenhouse gas 'Plan B' gaining support*, Fred Pearce, New Scientist, 10th December 2003. http://www.newscientist.com/news/news.jsp?id=ns99994467
Observer, 2003	*Bush Covers Up Climate Research*, The Observer, 21st September

2003. http://observer.guardian.co.uk/international/story/
0,6903,1046363,00.html

Observer 2004a   *Oil Addicts – it lubricates our lives, but can we do without the black stuff*, Heather Stewart, The Observer, Sunday 22nd August 2004. http://www.guardian.co.uk/oil/story/0,11319,1288178,00.html

Observer, 2004b   *Now the Pentagon Tells Bush: Climate Change Will Destroy Us*, The Observer, 22nd February 2004. http://www.guardian.co.uk/climatechange/story/0,12374,1153530,00.html

Observer, 2004c   *Solar Power for all New Houses*, Nick Mathiason, The Observer, Sunday 3rd October 2004. http://observer.guardian.co.uk/business/story/0,6903,1318304,00.html

OECD, 1999a   *World Energy Prospects to 2020: Issues and Uncertainties*, Jean-Marie Bourdaire, IEA – published in *Energy: The Next Fifty Years*, OECD, 1999. http://www.oecd.org/dataoecd/37/55/ 17738498.pdf

OECD, 1999b   *Towards a Sustainable Energy Future*, Dieter M. Imboden (Swiss Federal Institute of Technology) and Carlo C. Jaeger (Swiss Federal institute of Environmental Science and Technology) – published in *Energy: The Next Fifty Years*, OECD, 1999. http://www.oecd.org/dataoecd/37/55/17738498.pdf

Öko-Institut, 1997   *Comparing Greenhouse-Gas Emissions and Abatement Costs of Nuclear and Alternative Energy Options from a Life-Cycle Perspective*, German Institute for Applied Ecology, paper presented to the CNIC Conference on Nuclear Energy and Greenhouse-Gas Emissions, Tokyo, November 1997. http://www.oeko.de/service/gemis/files/info/nuke_co2_en.pdf

Oswald, 2004   *The Arithmetic of Renewable Energy*, Andrew Oswald, Professor of Economics, University of Warwick, Jim Oswald, Energy Consultant, October 2004. http://www2.warwick.ac.uk/fac/soc/economics/staff/faculty/oswald/windaccountancy04.pdf

OU, 2003   *Energy Systems and Sustainability – Power for a Sustainable Future*, edited by Godfrey Boyle, Bob Everett and Janet Ramage for The Open University, Oxford University Press, 2003. http://eeru.open.ac.uk/index.htm

OU, 2004   *Renewable Energy – Power for a Sustainable Future*, edited by Godfrey Boyle, The Open University, Oxford University Press, 2004. http://eeru.open.ac.uk/index.htm

PIU, 2002   *The Energy Review*, Downing Street Performance and Innovation Unit Report, February 2002. http://www.number-10.gov.uk/su/energy/1.html

Pointcarbon, 2004   For recent reports on carbon trading see http://www.pointcar-bon.com/

POST, 2001   *UK Electricity Networks*, POST Note No.163, Parliamentary Office of Science and Technology, October 2001. http://www.parliament.uk/post/pn163.pdf

POST, 2002   *Prospects for a Hydrogen Economy*, POST Note No.186, Parliamentary Office of Science and Technology, October 2002. http://www.parliament.uk/post/pn186.pdf

POST, 2003   *Security of Electricity Supplies*, POST Note No.203, Parliamentary Office of Science and Technology, September 2003. http://www.parliament.uk/post/pn203.pdf

POST 2004   *The Future of UK Gas Supplies*, POST Note No. 230, Parliamentary Office of Science and Technology, October 2004. http://www.parliament.uk/documents/upload/postpn230.pdf

RCEP, 1976   *Nuclear Power and the Environment*, Sixth Report of the Royal Commission on Environmental Pollution, 1976.

RECP, 2000   *Energy – The Changing Climate*, Twenty-Second Report of the Royal Commission on Environmental Pollution (Cm4749), June 2000. http://www.rcep.org.uk/newenergy.htm

Ricardo, 2002   *Carbon to Hydrocarbon Road Maps for Passenger Cars: A Study for the Department of Transport and the Department of Trade and Industry*, Ricardo Consulting Engineers under contract to the Department for Transport, November 2002. http://www.dft.gov.uk/stellent/groups/dft_roads/documents/page/dft_roads_507528.pdf

Rowell, 1996   *Green Backlash – Global Subversion of the Environment Movement*, Routledge, 1996. http://www.andyrowell.com/

SAFE, 1998   *The Perfect Pinta*, The SAFE Alliance, 1998. http://www.sustainweb.org/ffact_pinta.asp

Schumacher, 1973   *Small is Beautiful*, E.F. Schumacher, Blond and Briggs 1973. http://www.schumacher.org.uk/

Seymour, 1973   *Self-sufficiency – The Science and Art of Producing and Preserving Your Own Food*, John and Sally Seymour, Faber and Faber 1973 (revised and re-issued as by John Seymour as *The New Book of Complete Self-Sufficiency* in 2002). http://www.self-sufficiency.net/

Solarex, 1996   *World Design Insolation*, Solarex 1996. Available via BP Solar's web site. http://www.bpsolar.com/ContentDocuments %5C17% 5CPV%20System%20Sizing%20Tools.zip

Stauber, 1995   *Toxic Sludge is Good For You – Lies, Damn Lies, and the Public Relations Industry*, John Stauber and Sheldon Newman, USA Common Courage Press, 1995 (UK edition published by Robinson, 2004).

Sustain, 2001   *Eating Oil: Food Supply and Climate Change*, Sustain and Elm Farm Research Centre, 2001. http://www.sustainweb.org/ chain_fm_index.asp

Tickell, 1998   *From the Fryer to the Fuel Tank – The Complete Guide to Using Vegetable Oil as and Alternative Fuel*, Joshua and Kaia Tickell, Tickell Energy Consulting, 1998. http://www.joshuatickell.com/

Tidal Electric, 2004   *Feasibility Study for a Tidal Lagoon in Swansea Bay – Executive Summary*, Atkins Consultants for Tidal Electric Ltd., September 2004. http://www.tidalelectric.com/Web%20Atkins%20 Executive%20Summary.htm

Times, 2004   *Fears over reserves send Shell lower*, The Times, 9th January 2004. http://business.timesonline.co.uk/article/ 0,,8903-957359,00.html

UKT, 2000   Annex A, *Pre-Budget Report 2000*, UK Treasury. http://www.hm-treasury.gov.uk/pre_budget_report/pre_budget_report_2000/ pbr_report/prebud_pbr00_repannexa.cfm

UNCED, 1992   *The Rio Declaration on Environment and Development*, Second United Nations Conference on Environment and Development 1992. http://www.un.org/documents/ga/conf151/aconf15126-1annex1.htm

UNDP, 2000   *World Energy Assessment – Energy and the Challenge of Sustainability*, United Nation Development Programme (UNDP), 2000 http://www.undp.org/seed/eap/activities/wea/drafts- frame.html

UNEP, 2002   *Global Environmental Outlook 3*, United Nations Development Programme, 2002 (published by Earthscan Books).

USDOE, 2004   *Annual Energy Outlook 2004 with Projections to 2025*, DOE/EIA-0383(2004), US Department of Energy Energy Information Administration, January 2004. http://www.eia.doe.gov/ oiaf/aeo/index.html

USEPA, 2003   Issue Paper: *Whitehouse Edits to Climate Change Section of EPA's Report on the Environment*, US EPA (leaked) internal memo, 29th April 2003. http://www.fraw.org.uk/ebo/mirror/Issue-Paper.pdf

USGS, 2000   *World Petroleum Assessment 2000*, United Stated Geological Survey, 2000. http://pubs.usgs.gov/dds/dds-060/

Verne, 1874   Part 2/Chapter 11, *The Mysterious Island*, Jules Verne, 1874. http://www.mastertexts.com/Verne_Jules/The_Mysterious_

Island/Index.htm

Warwick, 2004   *Researchers Say Hydrogen Powered Cars Would Need 100,000 Wind Turbines or 100 Nuclear Plants*, University of Warwick Press Release 134, 6th October 2004. http://www2.warwick.ac.uk/ newsandevents/pressreleases/NE100000009439/

Ward, 1972   *Only One Earth – The Care and Maintenance of a Small Planet*, Barbara Ward and René Dubos, André Deutsch 1972.

Weizsäcker, 1998   *Factor Four: Doubling Wealth, Halving Resource Use*, Ernst von Weizsäcker, Amory B. Lovins and L. Hunter Lovins, Earthscan Books 1998.

White, 1994   *Integrated Solid Waste Management: A Lifecycle Inventory*, Dr P. White, Dr M. Franke and P. Hindle (Procter and Gamble UK, Germany and Belgium), Blackie Academic and Professional 1994.

WHOI, 2000   *Iron Fertilization in Southern Ocean Increased Growth of Algae that Absorb Greenhouse Gases, and Could Cool Climate*, Woods Hole Oceanographic Institute, 12th October 2000. http://www.whoi.edu/media/iron.html

WHOI, 2003   *Abrupt Climate Change: Should We Be Worried?*, Robert B. Gagosian, Woods-Hole Oceanographic Institute, prepared for a panel on abrupt climate change at the World Economic Forum, January 2003. http://www.whoi.edu/institutes/occi/currenttopics/climate-change_wef.html

Wired, 2004   *Hybrid Mileage Comes Up Short*, John Gatner, Wired News, 11th May 2004. http://www.wired.com/news/autotech/ 0,2554,63413,00.html

WP, 2003   *How Bush and Co. Obscured the Science*, Washington Post, 13th July 2003. http://www.washingtonpost.com/ac2/wp-dyn/ A46181-2003Jul11

Further Reading

Energy may be a complex issue, but the collation and use of energy statistics can sometimes appear to be one of the black arts. It's very difficult to find studies that provide detailed explanations as to how information was produced, or to what extent the assumptions within the energy models of governments and consultants are based upon real data or guess-work. If the content of this book has stirred an interest in how the energy economy works then I would recommend the following web sites.

First and foremost, the web site that goes with the book. This contains free information, information on upcoming events associated with the book, and updates to the book's content and references:

> http://www.fraw.org.uk/ebo/

For the UK, the first port of call for energy information is the *Department of Trade and Industry*. They are responsible for energy policy, and collate the UK's official energy statistics. Via the energy section of the DTI's website you can get the latest information on the UK's conventional and renewable energy resources – http://www.dti.gov.uk/energy/

The major source of statistical data on energy in the UK is the DTI's *Digest of UK Energy Statistics* (DUKES):

> http://www.dti.gov.uk/energy/inform/dukes/index.shtml

Tracking energy stories in the UK media is a little more difficult. Probably the most in-depth coverage is provided by the Financial Times (http://www.ft.com/), but for a quick review of energy stories currently in the media the easiest to digest are The Guardian's special reports pages: Oil and Petrol – http://www.guardian.co.uk/oil/0,11319,608464,00.html Renewables – http://www.guardian.co.uk/renewable/0,2759,180749,00.html Climate Change –

> http://www.guardian.co.uk/climatechange/0,12374,782494,00.html

For global energy statistics one of the simplest and best presented publications is BP's Digest of World Energy Statistics:

> http://www.bp.com/subsection.do?categoryId=95&contentId=2006480

For more in-depth coverage on international energy issues the two most useful sites, which also link to many other relevant sites around the world, are the Organisation for Economic Co-operation and Development (OECD) and the International Energy Agency (IEA). However, the IEA have a habit of putting reports online once they're a couple of years out-of-date, and charging a hefty fee for access to the newer ones:

OECD: http://www.oecd.org/

IEA: http://www.iea.org/

Finding information on Peak Oil from mainstream sources is rather difficult – it's not something that they openly talk about. Instead there are a growing number of sites on the web devoted to the issue. Rather than accepting any of the information at face value, it's always a good idea to read the background material, or do a search for the relevant research, in order to check where the proprietor of the site is coming from (in terms of their approach – pro-coal, pro-nuclear, etc.). For more information do a search for "Peak Oil" using a search engine, or go to the following sites:

Hubbert's Peak of Oil Production: http://www.hubbertpeak.com/

The Association for the Study of
Peak Oil and Gas: http://www.peakoil.net/

Finally, thanks must go to the Free Range Network for their help and support. In particular for asking me to answer a troubling question in 2003, *how much energy is there left?* This book is the direct result of that question, but more specifically the workshops I organised with the The Earth Cymru Network on the issue of energy economics and renewable energy. For more information on current energy issues in Wales see their web site at:

http://www.fraw.org.uk/ecn/

Index

License

THE WORK (AS DEFINED BELOW) IS PROVIDED UNDER THE TERMS OF THIS CREATIVE COMMONS PUBLIC LICENSE ("CCPL" OR "LICENSE"). THE WORK IS PROTECTED BY COPYRIGHT AND/OR OTHER APPLICABLE LAW. ANY USE OF THE WORK OTHER THAN AS AUTHORIZED UNDER THIS LICENSE IS PROHIBITED.

BY EXERCISING ANY RIGHTS TO THE WORK PROVIDED HERE, YOU ACCEPT AND AGREE TO BE BOUND BY THE TERMS OF THIS LICENSE. THE LICENSOR GRANTS YOU THE RIGHTS CONTAINED HERE IN CONSIDERATION OF YOUR ACCEPTANCE OF SUCH TERMS AND CONDITIONS.

1. Definitions

"Collective Work" means a work, such as a periodical issue, anthology or encyclopaedia, in which the Work in its entirety in unmodified form, along with a number of other contributions, constituting separate and independent works in themselves, are assembled into a collective whole. A work that constitutes a Collective Work will not be considered a Derivative Work (as defined below) for the purposes of this License.

"Derivative Work" means a work based upon the Work or upon the Work and other pre-existing works, such as a translation, musical arrangement, dramatization, fictionalization, motion picture version, sound recording, art reproduction, abridgement, condensation, or any other form in which the Work may be recast, transformed, or adapted, except that a work that constitutes a Collective Work will not be considered a Derivative Work for the purpose of this License.

"Licensor" means the individual or entity that offers the Work under the terms of this License.

"Original Author" means the individual or entity who created the Work.

"Work" means the copyrightable work of authorship offered under the terms of this License.

"You" means an individual or entity exercising rights under this License who has not previously violated the terms of this License with respect to the Work, or who has received express permission from the Licensor to exercise rights under this License despite a previous violation.

2. Fair Use Rights

Nothing in this license is intended to reduce, limit, or restrict any rights arising from fair use, first sale or other limitations on the exclusive rights of the copyright owner under copyright law or other applicable laws.

3. License Grant

Subject to the terms and conditions of this License, Licensor hereby grants You a worldwide, royalty-free, non-exclusive, perpetual (for the duration of the applicable copyright) license to exercise the rights in the Work as stated below:

to reproduce the Work, to incorporate the Work into one or more Collective Works, and to reproduce the Work as incorporated in the Collective Works;

to create and reproduce Derivative Works;

to distribute copies or phono-records of, display publicly, perform publicly, and perform publicly by means of a digital audio transmission the Work including as incorporated in Collective Works;

to distribute copies or phono-records of, display publicly, perform publicly, and perform publicly by means of a digital audio transmission Derivative Works;

The above rights may be exercised in all media and formats whether now known or hereafter devised. The above rights include the right to make such modifications as are technically necessary to exercise the rights in other media and formats. All rights not expressly granted by Licensor are hereby reserved.

<u>4. Restrictions</u>

The license granted in Section 3 above is expressly made subject to and limited by the following restrictions:

You may distribute, publicly display, publicly perform, or publicly digitally perform the Work only under the terms of this License, and You must include a copy of, or the Uniform Resource Identifier for, this License with every copy or phono-record of the Work You distribute, publicly display, publicly perform, or publicly digitally perform. You may not offer or impose any terms on the Work that alter or restrict the terms of this License or the recipients' exercise of the rights granted hereunder. You may not sub-license the Work. You must keep intact all notices that refer to this License and to the disclaimer of warranties. You may not distribute, publicly display, publicly perform, or publicly digitally perform the Work with any technological measures that control access or use of the Work in a manner inconsistent with the terms of this License Agreement. The above applies to the Work as incorporated in a Collective Work, but this does not require the Collective Work apart from the Work itself to be made subject to the terms of this License. If You create a Collective Work, upon notice from any Licensor You must, to the extent practicable, remove from the Collective Work any reference to such Licensor or the Original Author, as requested. If You create a Derivative Work, upon notice from any Licensor You must, to the extent practicable, remove from the Derivative Work any reference to such Licensor or the Original Author, as requested.

You may distribute, publicly display, publicly perform, or publicly digitally perform a Derivative Work only under the terms of this License, and You must include a copy of, or the Uniform Resource Identifier for, this License with every copy or phono-record of each Derivative Work You distribute, publicly display, publicly perform, or publicly digitally perform. You may not offer or impose any terms on the Derivative Works that alter or restrict the terms of this License or the recipients' exercise of the rights granted hereunder, and You must keep intact all notices that refer to this License and to the disclaimer of warranties. You may not distribute, publicly display, publicly perform, or publicly digitally perform the Derivative Work with any technological measures that control access or use of the Work in a manner inconsistent with the terms of this License Agreement. The above applies to the Derivative Work as incorporated in a Collective Work, but this does not require the Collective Work apart from the Derivative Work itself to be made subject to the terms of this License.

You may not exercise any of the rights granted to You in Section 3 above in any manner that is primarily intended for or directed toward commercial advantage or private monetary compensation. The exchange of the Work for other copyrighted works by means of digital file-sharing or otherwise shall not be considered to be intended for or directed toward commercial advantage or private monetary compensation, provided there is no payment of any monetary compensation in connection with the exchange of copyrighted works.

If you distribute, publicly display, publicly perform, or publicly digitally perform the Work or any Derivative Works or Collective Works, You must keep intact all copyright notices for the Work and give the Original Author credit reasonable to the medium or means You are utilizing by conveying the name (or pseudonym if applicable) of the Original Author if supplied; the title of the Work if supplied; in the case of a Derivative Work, a credit identifying the use of the Work in the Derivative Work (e.g., "French translation of the Work by Original Author," or "Screenplay based on original Work by Original Author"). Such credit may be implemented in any reasonable manner; provided, however, that in the case of a Derivative Work or Collective Work, at a minimum such credit will appear where any other comparable authorship credit appears and in a manner at least as prominent as such other comparable authorship credit.

5. Representations, Warranties and Disclaimer

By offering the Work for public release under this License, Licensor represents and warrants that, to the best of Licensor's knowledge after reasonable inquiry:

Licensor has secured all rights in the Work necessary to grant the license rights hereunder and to permit the lawful exercise of the rights granted hereunder without You having any obligation to pay any royalties, compulsory license fees, residuals or any other payments;

The Work does not infringe the copyright, trademark, publicity rights, common law rights or any other right of any third party or constitute defamation, invasion of privacy or other tortuous injury to any third party. EXCEPT AS EXPRESSLY STATED IN THIS LICENSE OR OTHERWISE

AGREED IN WRITING OR REQUIRED BY APPLICABLE LAW, THE WORK IS LICENSED ON AN "AS IS" BASIS, WITHOUT WARRANTIES OF ANY KIND, EITHER EXPRESS OR IMPLIED INCLUDING, WITHOUT LIMITATION, ANY WARRANTIES REGARDING THE CONTENTS OR ACCURACY OF THE WORK.

6. Limitation on Liability

EXCEPT TO THE EXTENT REQUIRED BY APPLICABLE LAW, AND EXCEPT FOR DAMAGES ARISING FROM LIABILITY TO A THIRD PARTY RESULTING FROM BREACH OF THE WARRANTIES IN SECTION 5, IN NO EVENT WILL LICENSOR BE LIABLE TO YOU ON ANY LEGAL THEORY FOR ANY SPECIAL, INCIDENTAL, CONSEQUENTIAL, PUNITIVE OR EXEMPLARY DAMAGES ARISING OUT OF THIS LICENSE OR THE USE OF THE WORK, EVEN IF LICENSOR HAS BEEN ADVISED OF THE POSSIBILITY OF SUCH DAMAGES.

7. Termination

This License and the rights granted hereunder will terminate automatically upon any breach by You of the terms of this License. Individuals or entities who have received Derivative Works or Collective Works from You under this License, however, will not have their licenses terminated provided such individuals or entities remain in full compliance with those licenses. Sections 1, 2, 5, 6, 7, and 8 will survive any termination of this License.

Subject to the above terms and conditions, the license granted here is perpetual (for the duration of the applicable copyright in the Work). Notwithstanding the above, Licensor reserves the right to release the Work under different license terms or to stop distributing the Work at any time; provided, however that any such election will not serve to withdraw this License (or any other license that has been, or is required to be, granted under the terms of this License), and this License will continue in full force and effect unless terminated as stated above.

8. Miscellaneous

Each time You distribute or publicly digitally perform the Work or a Collective Work, the Licensor offers to the recipient a license to the Work on the same terms and conditions as the license granted to You under this License.

Each time You distribute or publicly digitally perform a Derivative Work, Licensor offers to the recipient a license to the original Work on the same terms and conditions as the license granted to You under this License.

If any provision of this License is invalid or unenforceable under applicable law, it shall not affect the validity or enforceability of the remainder of the terms of this License, and without further action by the parties to this agreement, such provision shall be reformed to the minimum extent necessary to make such provision valid and enforceable.

No term or provision of this License shall be deemed waived and no breach consented to unless such waiver or consent shall be in writing and signed by the party to be charged with such waiver or consent.

This License constitutes the entire agreement between the parties with respect to the Work licensed here. There are no understandings, agreements or representations with respect to the Work not specified here. Licensor shall not be bound by any additional provisions that may appear in any communication from You. This License may not be modified without the mutual written agreement of the Licensor and You.